Micro Irrigation Technician

Micro Irrigation Technician

Rajesh S
Sreejith K

2026
Agri Horti Press
New Delhi-110 084

ISBN: 978-93-49690-28-8 (Hardbound)

Published By

Agri Horti Press

New Delhi

E-mail: agrihortipress@gmail.com

Digitally Printed at : Replika Press Pvt. Ltd.

Preface

Micro Irrigation Technician serves as a comprehensive guide for understanding the principles, technologies, and practical applications of micro-irrigation systems. As the demand for efficient water management in agriculture grows due to climate change, water scarcity, and increasing global food demands, micro-irrigation technologies have emerged as a vital solution to optimize water use, improve crop yields, and promote sustainable farming practices.

This handbook is designed to provide aspiring technicians, agricultural professionals, and students with a detailed understanding of micro-irrigation systems, their design, installation, operation, and maintenance. The book explores a range of micro-irrigation methods, including drip irrigation, sprinkler irrigation, and subsurface drip irrigation, highlighting their advantages in water conservation, nutrient delivery, and crop management.

The first section of the book introduces the basic concepts of irrigation, including the importance of water management in agriculture and the principles of soil-water-plant relationships. The text then delves into the specific components and technologies involved in micro-irrigation systems, such as pumps, filters, emitters, controllers, and piping systems. Each component is thoroughly explained, with an emphasis on its function, design considerations, and selection criteria based on crop type, field conditions, and water availability.

In addition to the technical aspects of system design, the book provides practical guidance on the installation of micro-irrigation systems. Detailed step-by-step procedures are provided for the setup of drip and sprinkler irrigation systems, along with troubleshooting tips to address common issues such as clogs, pressure imbalances, and system inefficiencies. The book also covers advanced topics such as automation, remote monitoring, and integration of micro-irrigation systems with precision farming technologies.

The volume also highlights the economic and environmental benefits of micro-irrigation, emphasizing its role in reducing water usage, lowering energy consumption, and enhancing crop production. Case studies and real-world examples are incorporated to illustrate the successful implementation of micro-irrigation in various agricultural settings, ranging from small-scale farms to large commercial operations.

Sustainability is a central theme throughout the book. In addition to providing practical knowledge for technicians, the text encourages a holistic understanding of how micro-irrigation systems can contribute to long-term water conservation, improved soil health, and enhanced agricultural productivity in the face of environmental challenges.

The *Micro Irrigation Technician* handbook is an essential resource for anyone involved in the design, installation, and maintenance of micro-irrigation systems. Whether you are a technician, farmer, researcher, or educator, this book will provide the knowledge and tools necessary to apply micro-irrigation techniques effectively and efficiently.

We extend our gratitude to the experts, engineers, and agricultural professionals whose knowledge and contributions have informed the development of this volume. It is our hope that this book serves as both a practical guide and an educational resource to further the adoption of sustainable irrigation practices worldwide.

Rajesh S

Sreejith K

Contents

Chapter 1

Introduction

1.1. Scope of Micro Irrigation in India

India is an agrarian society where agricultural sector accounts for 20% of India's gross domestic product giving employment to 50% of country's workforce. India initiated its own Green Revolution programme in plant breeding, irrigation development and financing of agrochemicals after 1960.

The Green Revolution had several benefits like increase in production due to use of high-yielding varieties of seeds, crop genetic improvements and irrigation, which led to widespread poverty reduction. The unabated adoption of unsustainable agricultural techniques and practices by farmers to produce more had an adverse impact on the environment.

National Institution for Transforming India in 2018 stated that around 600 million Indians are now facing high to extreme water stress situation due to growing population and ever-increasing demand for food. High variations in range of electrical conductivity between 2.85 and 6.0 ds/m and heavy metals concentration in groundwater shows high salinity problem that exacerbated by groundwater withdrawals and contamination of groundwater.

The average annual water availability in India is estimated to be 1869 billion cubic meters, but due to hydrological, topographic and other constraints, usable water is expected to be about 1123 BCM, out of which 690 BCM is from surface water and 433 BCM from replenishable groundwater.

The water use efficiency in Indian agriculture is around 30-40%, which is one of the lowest in the world, so judicious use of irrigation water requires more attention to upgrade total agricultural production and area under irrigated agriculture. This can be done by using advance method of irrigation like micro-irrigation which can be coupled with other improved management practices.

The increasing demand of water from other sectors with inefficient methods of irrigation has aggravated the problems of water scarcity, so to tackle water scarcity problem in India in agriculture sector, Government of India has come up with many irrigation programmes and schemes from time to time. In early times, farmers were based on open canal system concept, but with change in current and upcoming irrigation programmes and schemes from canal flood irrigation to modern micro irrigation system that uses drip and sprinkler irrigation methods.

The micro irrigation systems comprises of:

- ❖ drip irrigation system
- ❖ sprinkler irrigation system
- ❖ micro-sprinkler irrigation system
- ❖ mini-sprinkler irrigation system

As seen, drip and sprinkler irrigation system methods of irrigation helps in saving water and improves WP as drip and sprinkler methods of irrigation are treated as micro irrigation system, so there are distinct characteristic differences between the two in terms of flow rate, pressure requirement, wetted area and mobility.

Indian micro irrigation system industry, which comprises of segments like drip irrigation and sprinkler irrigation has grown significantly over last few years due to increase in demand for agricultural food products on account of burgeoning population in the country. The micro irrigation system industry in India is highly fragmented with nearly 200 companies. There is a fierce competition in the market with a large number of organized and unorganized players engaged in the manufacturing and installation of micro irrigation system in the industry.

There is a huge scope for micro irrigation systems and many micro irrigation schemes offer subsides ranging from 50% to 95%. There have been various success stories and projects on community-based approach for supplying irrigation water to agricultural areas that were earlier under rainfed agriculture and faced crop failures due to scanty rainfall showcasing the benefits of increased crop production and reduced cost of production with minimal water and power consumption through micro irrigation method adoption.

Today, India is considered as water stressed country, which were highly surplus in comparison with 1950s. Such change appears due to:

- ❖ increasing population
- ❖ highly wasteful flood method of irrigation

As around 65% of irrigation and 85% if drinking water comes from groundwater resources, majority of groundwater sources in the country are registering a fall in water levels. To arrest the decline in per capita water availability and per capita food grain availability, an increase in production with judicious water use in farming is required.

Micro irrigation importance also increases given that 140 million hectares is owned millions of farmers with an average farm size of less than two hectares. Such irrigation systems not only address the problem of water scarcity but also effectively save fertilizer consumption per unit of land. Due to micro irrigation system, nutrients are released through the system, rather than their application on top of soil which leads to wastage during the periodic flooding of fields under FMI.

The governments across the world support micro-irrigation due to its water use efficiency. The government of India had launched a subsidy scheme for micro-irrigation called National Mission on Micro Irrigation in 2006 which is a centrally sponsored scheme that promotes water-use efficiency by adopting drip and sprinkler irrigation systems where all States and Union Territories and all horticultural and agricultural crops are covered under this scheme.

1.2 Benefits for adopting micro irrigation

Micro irrigation allows conservation and efficient use of water, less water wastage and higher crop productivity with less water consumption using drip and sprinkler irrigation methods. For optimal and efficient use of surface and groundwater sources for irrigation, micro irrigation method usage is one of the effective ways to grow more crops with less water.

Micro irrigation could be one of the solutions to the water resources management challenges and issues faced by Indian agriculture. The water use efficiency of flood method of irrigation in India is estimated to be only around 40% which is due to significant losses by conveyance, distribution and evaporation, while micro irrigation systems provide water use efficiency from 80% to 95%.

Micro irrigation systems are quite efficient as they have nominal transmission losses, while losses through evaporation, run-off and deep percolation are also reduced significantly by using them. Efficient water use results in additional benefits such as an increase in the area coverage under irrigation with the same amount of water as well as increasing the potential usage of marginal/degraded land using micro irrigation systems.

There are several benefits for adopting micro irrigation:

- ❖ Efficient increase in water usage
- ❖ Reduction in energy consumption
- ❖ Reduction in fertiliser consumption
- ❖ Productivity upgrade of fruits/crops and vegetables
- ❖ Saving in irrigation cost
- ❖ Increasing farmers' income

1.3 Features of micro irrigation system

- ❖ Water is applied via pressurized piping system. Micro irrigation requires pumps for developing the required pressure for delivering water through pipelines, regardless of whether the source of water is surface or underground.
- ❖ Water is applied drop-by-drop for a long period in case of drip irrigation system.

- Water is applied at a low rate to maintain the optimum air–water balance within the root zone.
- Water is applied at frequent intervals as per the requirement of plants.
- Water is supplied directly to the plants and not to the other areas of the field, thus, reducing wastage.
- Soil moisture content is always maintained at 'field capacity' of the soil. Hence, crops grow at a faster rate, consistently and uniformly.

1.4 Advantages of Micro Irrigation Systems

Some of the major advantages of Micro Irrigation System are given below:

- Affordability: Micro Irrigation System is available in affordable sizes from local suppliers at prices lower than other available irrigation systems.
- Improved Yield: Slow and regular application of water and nutrients uniformly to all plants improves product quality and uniformity, and increases yield.
- Water Saving: In this water savings is around 50% as compared to traditional irrigation methods. Using such system , a farmer can irrigate more crop area per unit of water used.
- Labour Saving: In this, less labor is required for irrigation, weeding, and fertilizer application as compared to traditional production methods.
- Fertilizer Saving: In such system, a fertilizer loss is minimized with less fertilizer costs.
- Energy Saving: Many such system are gravity operated or can operate with low horse power pumps, reducing energy demand for irrigation.
- Difficult Terrain: Micro Irrigation System can be used on undulated terrain whereas irrigation by traditional methods is difficult.
- Tolerance to Salinity: Due to slow and regular application of water by such irrigation system, concentration of salts in the root zone is reduced and by micro-leaching salts are kept away from root zone.
- Improved Crop and Disease Control: Regular irrigation allows timely inter-culturing operations and spraying which gives better control over potential crop diseases which reduces incidence of diseases common with flood irrigation.
- Reduced Cultivation Cost: Slow and regular application of water keeps optimum soil-water-air ratio in soil which is required for healthy plant growth. It also reduces the need for frequent inter-culturing, weeding, etc. Combined with the above-mentioned savings, cultivation costs on the whole are reduced.
- Application to Variety of Crops: A number of different crops can be irrigated using such irrigation system which includes vegetable crops, fruit crops, commercial cash crops, flowers, etc.

1.5 Issues/Challenges of Micro Irrigation for improving water Usage

Apart from benefits of modern irrigation methods like surface or sub-surface drip and sprinkler irrigation systems, such systems are not commonly adopted on large scale by farmers due to:

- high capital cost
- non-reliable sources of water for system
- marginal and scattered agricultural landholdings
- unavailability of subsidy at required time
- delayed release of funds for installation which was already approved
- absence of easy financing mechanisms
- lack of knowledge

As a result of periodic changes in components/schemes related to National Mission on Micro Irrigation, National Mission for Sustainable Agriculture and Prime Minister Krishi Sinchayee Yojana, many stakeholders have stated that there seems to be lack of focus on micro irrigation. With literature survey and consultation with stakeholders, there are certain major issues that hindered the growth of micro irrigation, which includes:

1.5.1. Tedious financing machinery to beneficiaries

Financing for beneficiaries continues to be a major obstacle, as a result of which they continuously face difficulties in securing financing options. Through consultation with stakeholders it was found that there is a persistent problem with the unavailability of funds in some states, resulting in further delay in fund disbursement from the state government to the dedicated cell/officer responsible for promoting and initiating MIS schemes at the state level. This in turn leads to lengthy cycles of subsidy. There is a necessity to find ways to guarantee availability of funds on time and for simpler funding norms for farmers on priority.

1.5.2. Insecurity and periodic variations in irrigation scheme strategies

Owing to the lack of effective and long-term strategies, most of the schemes are operational only for a limited period of time at the state level. In this context, Federation of Indian Chambers of Commerce and Industry report has revealed that on average schemes are operative only for five months in a year to the farmers. As a result, farmers could not avail the benefits of the scheme during peak water demand months. Also, this could result in the farmers missing the cropping season, leading to their inability to gauge the real benefits of the micro irrigation system.

1.5.3. Scheme execution and functioning processes

Most of the micro irrigation schemes are inefficient in terms of their execution and functioning, particularly in terms of the time taken and length of the process of scheme execution. In a few states, the scheme implementation and launch process occurs during the months of August and September, as a result of which there is a lag in irrigation water supply and timely irrigation to the field. By the time a scheme actually starts to provide water for crop, the main season has already ended. Therefore, beneficiaries are not able to get the optimal benefit of the available micro irrigation technology. Thus, there is a need for timely completion of the process for better synchronisation with the farmers' interest and demands. The lack of a robust monitoring system makes the task of managing scheme execution and functioning extremely difficult. This has a detrimental impact on the efficiency of implementing MIS.

1.5.4. Lack of motivation on micro irrigation

The perception of a diluted focus on the micro irrigation scheme is still prevalent, with the Government of India providing funds for MIS since 2014–15 as a component of government schemes and not through the dedicated mission such as the previous National Mission on Micro Irrigation. This is also followed to a large extent at the state level as only a few states have a dedicated team to promote micro irrigation, and most others do not provide the due importance needed, thus leading to deceleration in the growth of MIS.

1.6 Understand Role of Micro Irrigation Technician

Micro-irrigation involves in irrigation of crop at root zone as per crop need by appropriate technology which covers drip systems, sprinkler systems, etc. With respect to big improvement in water use efficiency, micro irrigation technology can also be used to facilitate optimal and effective application of fertilizers.

Micro Irrigation Technician is a person who has a very critical role in installation and function of micro irrigation system which not only uses the existing natural resources efficiently but also benefits the farmers to decrease the cost of cultivation in long run.

The work of the Micro Irrigation Technician is to install, test, and commission the micro irrigation system at field level and to make sure that uninterrupted supply of water to the plant stands on the farm.

1.6.1 Role of Micro Irrigation Technician

- ❖ To work independently and to comfortably make decisions in his/her area of work
- ❖ To be clear, skillfull to basic arithmetic and algebraic principles
- ❖ The person should be result oriented and is responsible for his/her own working and learning
- ❖ To demonstrate skills of using various tools and decision making for instant problem solving

1.7 Types of Micro Irrigation System

1.7.1 Micro-irrigation

Micro irrigation is the slow application of continuous drips, tiny streams or miniature sprays of water above or below the soil surface. It is used to cover all forms of small emission devices such as emitters, row crop tubing, spray strakes and micro-sprinklers.

Micro irrigation system is effective in saving water and increases water use efficiency as compared to conventional surface irrigation method. Such system will help in:

- ❖ reducing water consumption,

- growth of unwanted plants (weeds)
- soil erosion
- cost of cultivation

Micro irrigation can be adopted in all kinds of land, especially where it is not possible to effectively use flooding method for irrigation. In flooding method of irrigation, a field is flooded with water which results in significant run-off, anaerobic conditions in soil and around root zone, and deep irrigation below the root zone, which does not supply sufficient water to plants. It is one of the most inefficient surface irrigation methods.

Micro irrigation can be useful in undulating terrain, rolling topography, hilly areas, barren land and areas having shallow soils. According to depth, soil types can be classified as shallow, medium deep and deep soil.

1.7.2 Types

The micro irrigation systems comprises of:

- drip irrigation system
- sprinkler irrigation system
- micro-sprinkler irrigation system
- mini-sprinkler irrigation system

1. Drip irrigation system

Drip irrigation system is a method of applying required amount of water directly to the root zones of plants by using drippers or emitters at frequent intervals. In this, water is applied drop-by-drop or by micro jet on the soil surface or sub-surface at rate lower than the infiltration rate of the soil. In this, emitters dissipate pressure from the distribution system using orifices, vortexes and tortuous or long flow paths and allow limited amount of water to discharge. Many emitters are kept on the ground but they can also be buried. The emitted water moves in the soil system mainly by unsaturated flow. The water moves in the soil and wets the root zones of plants vertically by gravity and laterally by capillary action. In this, lateral movement of water beneath the surface is more in medium for heavy soil in comparison to sandy soil. The wetted soil area for widely spaced emitters is elliptical in shape. Such type of irrigation system can be used on windy days and during various land operations.

Types:

Drip irrigation system can be classified as:

(i) Surface drip irrigation

(ii) Sub-surface drip irrigation

(iii) Family drip

(iv) Online drip

(v) In-line drip

Surface drip irrigation

Surface drip irrigation is used to irrigate perennial and annual crops. Surface drip irrigation system is generally used to irrigate high-value vegetable crops like tomato, broccoli, celery, cauliflower, spinach, kohlrabi, leaf lettuce, etc.

Typical surface drip irrigation system consists of:

- **Pump unit:** It comprises of pump and power unit to supply electricity to pump. The pump draws water from the source and provides the right pressure for its delivery into the pipe system.
- **Head control unit:** It consists of shut-off, air and check valves to control the discharge and pressure of water in entire system. A pressure relief valve is installed after the pump unit to return excess water when the system is not operated at its full capacity. It may also have filters to clear the water. The filters remove sediment and debris, which can clog the system. Disc filters are commonly used to filter water from ponds, reservoirs, tanks and other sources that contain algae. Some head control units contain a fertiliser or nutrient tank to supply fertiliser solution to plants.
- **Tubings:** It consists of a main line, sub-main lines or sub-mains and laterals. The main line conveys water from the source and distributes it to the sub-mains. The sub-mains convey water to the laterals, which in turn supply it to the emmiters or drippers. The laterals are, usually, 13–32 mm in diameter and supply water into fields through the head control unit.
- **Emitters or drippers:** These devices are used to control the discharge of water from the laterals to plants. They are made of High Density Polythylene plastic. Water enters the drippers at approximately 1 kg/cm2 pressure and is delivered at zero pressure in the form of droplets at a low rate of 1–2.4 litre/hour. There are mainly two types of emitters.

(a) Online emitters:

- These are small plastic devices, which convey small streams of water from polyethylene tubing to the soil.
- In this, water moves through the soil by capillary flow and creates a wetted circle, the size of which depends on soil type, flow rate and irrigation schedule.
- Such emitters are attached to PE tubing wall by inserting emitters' barb-shaped base by punched hole.
- The emitters can be placed anywhere along the length of the pipe with emitters as self-piercing barbs.
- The diameter of pipes used for installing online emitters are usually between 12-20 mm.

(b) In-line emitters or drip lines:

These consist of small plastic emission devices, which are pre-inserted into the PE tubing at specified intervals during the tubing extrusion process.

Their rate of water flow depends on the inlet pressure.

With lower inlet pressure, the water flow decreases, whereas, with high pressure, it increases.

This emitter is available in 0.8 lph to 4 lph discharge rate.

Sub-surface drip irrigation

Subsurface drip irrigation is a variation from conventional surface drip irrigation techniques. By minimising evaporation it uses water more efficiently than surface irrigation. The depth that the laterals are buried at depends mostly on the tillage practices and the crop to be irrigated. It is a costly system that needs expert design and maintenance. But it is also a very effective method that transports the water directly to the roots and is especially appropriate for arid and windy locations. Before entering the distribution pipes the water is filtered to minimise the risk of clogging. Still, the system must be designed so that all pipes can be flushed periodically. As it requires high maintenance and expert design, subsurface drip irrigation systems are more suitable for medium and large-scale production.

Subsurface drip irrigation system has a similar design as common drip irrigation system. It consists of:

- settling pond,
- pumping unit,
- pressure relief valve,
- check valve or back flow prevention valves,
- hydrocyclone separator,
- chemical/fertiliser injection unit
- filtration unit equipped with back flush valves,
- pressure regulators,
- air vent valves
- PVC pipes delivering water to crop.

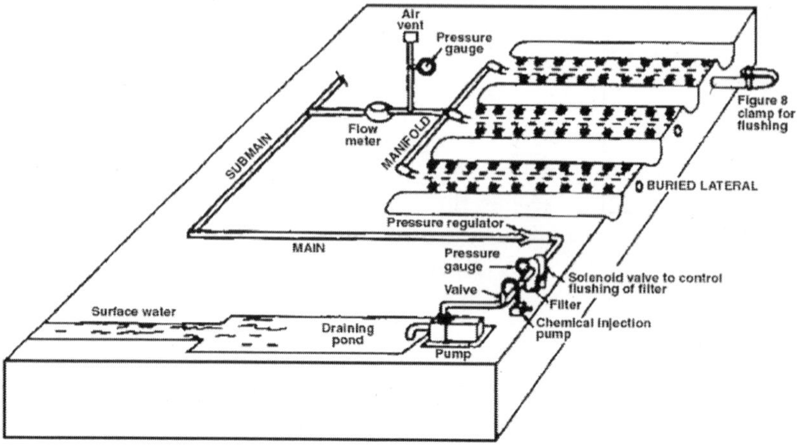

Sub-surface drip irrigation is a method of irrigating crops through buried plastic tubes, containing embedded emitters located at regular spacing's. A subsurface drip irrigation system has a similar design as surface drip irrigation system. But in this case, the drip tubes are typically located 38–84″ apart and 6–10″ below the soil surface. In sub-surface drip irrigation, evaporation is minimised and water is used more efficiently as compared to surface irrigation.

In sub-surface irrigation, the effects of surface infiltration like crusting, water losses via evaporation and surface run-off are eliminated. Water is applied directly to the root zone of a crop as opposed to surface irrigation, in which most weed seeds hibernate. Water application is efficient and uniform in this system. Sub-surface drip irrigation helps in water conservation in open field agriculture, often resulting in saving up to 25–50 percent water as compared to the flood irrigation system.

Subsurface drip irrigation systems are highly efficient irrigation systems that apply accurate amounts of water directly to the root zone, preventing water loss through evaporation and other negative effects of surface irrigation. This is especially suitable for arid, semi-arid, hot, and windy areas with limited water supply. However, as the system is relatively complex and most likely automated, it is more suitable for medium to large-scale production.

Advantages

- High degree of control over water application with the potential for high uniformity of application
- Evaporation is reduced
- The amount of water can be fine-tuned. This avoids water loss caused by run off or evaporation
- Frequent irrigation allows for optimum soil moisture content in the root zone
- Great performance in windy and arid locations

- ❖ If pre-treated wastewater is used for irrigation, the risk of direct contact with crops and labourers is reduced

Disadvantages

- ❖ Risk of clogging
- ❖ When saline water is used, salts accumulate at the wetting front
- ❖ Emitter can be damaged or blocked by root hairs
- ❖ Bacterial slimes and algae growing on the interior walls of the laterals and emitters combined with clay particles in the water can block the emitters
- ❖ Suspended organic matter and clay particles can damage the system
- ❖ A lot of repair work is caused by rodents chewing the tubes
- ❖ Heavy machinery can damage the laterals

Family drip irrigation

Family drip irrigation system is a low-cost system developed for small family plots. It is suitable for house gardening and peri-urban agriculture. It can also be used to demonstrate the working of drip irrigation system. Family drip system is designed for areas measuring 500–1000 m^2. It consists of:

- ❖ elevated tank,
- ❖ shut-off valve,
- ❖ filter,
- ❖ main line
- ❖ drip line

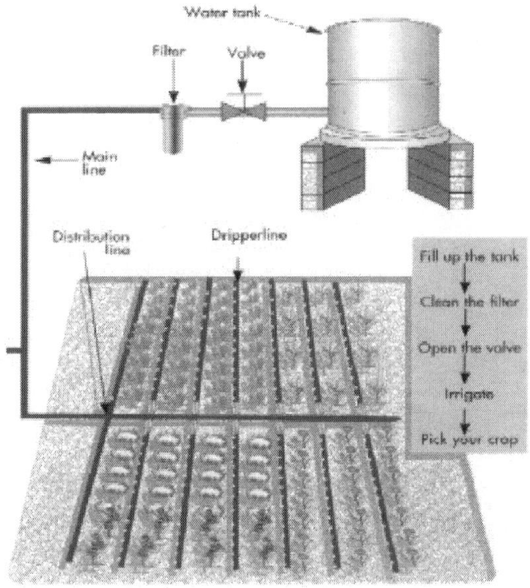

Family drip irrigation system comprises of drum, control or shut-off valve, filter, main line and drip laterals. The drip outlets are spaced at 30 cm. No central pressurised water system or power source is required in this system, hence such system is cheap, easy to install and operate.

Features:

- Easily attaches to any water tank
- Simple to install, operate and maintain
- No pump or electricity required for operation
- Applicable for a broad variety of water sources
- Effective for open field crops as well as small greenhouses
- Designed for all types of crops and for year-round application
- Suitable for plots of any shape or size: from 100 to 1 000 m2

Online drip irrigation

In this system, emitters or drippers are fixed externally on the laterals at designed spacings, so the drippers can be checked and cleaned easily in case of clogging.

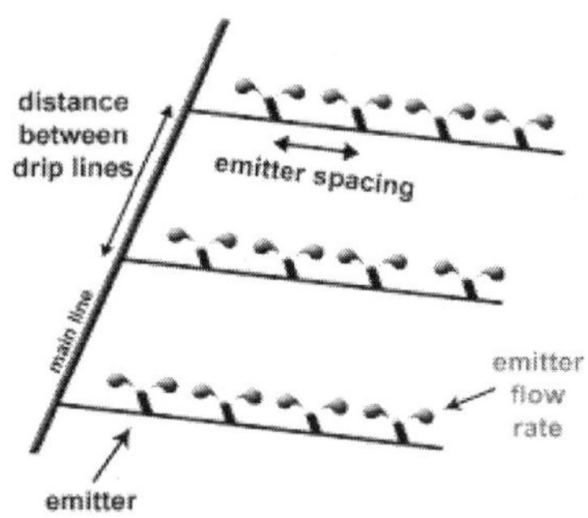

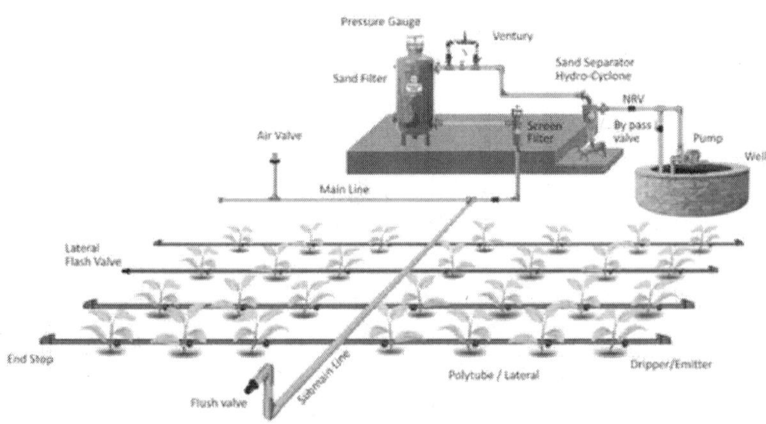

The dripper spacing can be changed any time to cover the increased root zone of a plant. Online dripper system is used in orchards, vineyards, artificial landscapes and nurseries.

It is, generally, used for irrigating horticultural plants like mango, coconut, orange, lemon, banana, grapes, pomegranate, papaya, sapota, guava, teakwood, bamboo, amla, etc.

In-line drip irrigation

In this system, drippers are fixed in the lateral tube at designed spacings at the time of manufacturing to meet the requirement of various crops.

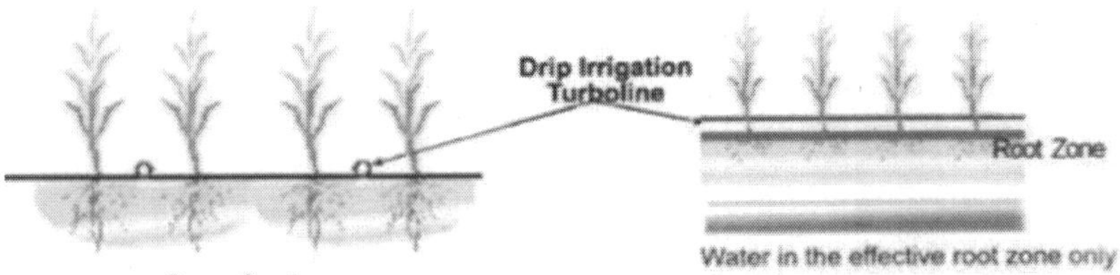

It is effective for row crops like cotton, sugarcane, groundnut, vegetables and flowering crops. Dripper spacing depends on the water requirement of a crop and the water holding capacity of the soil. Once installed, the dripper spacing cannot be changed.

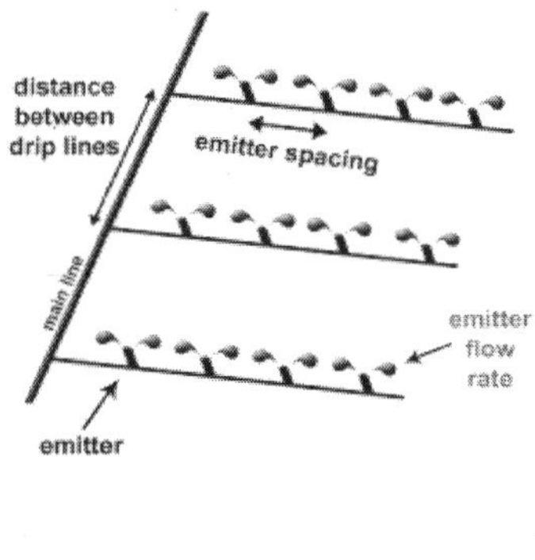

Components of inline drip system

Controller

It is the brain of every irrigation system which provide effective tools for irrigation management. For projects larger than 30 zones, use of two-wire decoder technology, available for both ACC and I-Core controllers are used. For areas without AC power, wall mount or in-valve-box battery controllers are ideal. All controllers are compatible with PLD, Eco-Wrap, and Eco-Mat.

ET-Sensor

ET-based controllers optimize scheduling based on real-time weather conditions to minimize water waste due to management factors. Solar Sync ET sensor adds even more efficiency to inline drip project by automatically adjusting irrigation scheduling for changing weather and site-specific climate conditions. The Solar Sync is compatible with all AC operated Hunter controllers. Both installation and operation are simple processes.

Valve

A remote control valve is typically used to automatically activate inline drip systems. A manual valve or hose bib may be used for non-automatic systems. Each hydrozone should be irrigated by a separate valve. Hydrozones are areas with specific conditions that affect irrigation, including plant type, spacing, density, microclimate, exposure, and slope.

Filter

Filtration must be provided for all inline drip systems, regardless of water supply type. Even potable water contains suspended particulates which may clog inline emitters. Filters also help reduce biological contaminants. For large or distributed systems, consider a disk filter near the point of connection which gives single, easily maintained item in accessible location. Use screen filters at each drip valve as a secondary level of protection in case any debris enters the irrigation system downstream of the primary filter.

Inline Drip Tubing

Inline drip tubing consists of tubing laid in parallel rows which creates a grid of emitters evenly spaced throughout the entire irrigated area.

Pressure Regulation

Basically all inline drip products has built-in pressure compensation that allows inlet pressure of 15-50 PSI. For mainline pressures exceeding 50 PSI, a pressure regulator downstream of valve and filter is used.

Laterals

Laterals moving water from valve to header(s) must be sized to accommodate the full flow of each area they serve. Industry standards are not to exceed a velocity of 5 FPS. Laterals are typically PE or PVC pipe. The may also be constructed of blank inline drip tubing for smaller areas.

Barbed Fittings

Barbed insert fittings gives positive connection to inline drip tubing. The insert fittings are colormatched with PLD which are UV resistant that handles pressure up to 200 PSI, and gives positive, easy-to-install, watertight connection without using tools, clamps, or glue.

Air/Vacuum Relief Valve

Inline drip emitters are designed to release air from system and block back siphonage, so AVR valves speed up the process and provides immediate path for large volumes of air to escape. Such can be installed on highest position of every contiguous area of inline drip tubing. AVR valves must be used when automatic flush valves are specified to prevent back-siphonage through emitters during the initial flush cycle.

Exhaust Header

The exhaust header serves to equalize pressure and flow between runs of inline drip tubing and provide an outlet path for flushing. The exhaust header does not need to be sized to equal the supply header, but must be able to accommodate the flow rate of the flush valve, without exceeding 5 FPS.

Flush Valves and PLD Cap

Flush valves are required on every inline drip system and must be located to provide an outlet from every point in each zone. Regular flushing not only removes debris and particles from the tubing, but the high volume of turbulent water during flushing also helps dislodge biological growth. Flush valves can be either automatic or manual.

Supply Header

Each supply header must be sized to accommodate full flow of entire area it serves. In this, headers may be either end-feed or center-feed configurations.

2. Sprinkler irrigation system

Introduction

Sprinkler irrigation is a method of applying water in a manner similar to rain. It is suited for most row, field and tree crops. Water can be sprayed over or under the crop canopy. If a site is known to be windy most of the time, sprinkler irrigation will not be suitable. The sprinkler breaks up the water into droplets sized 0.5–4 mm. The drop size is controlled by pressure and nozzle size of the sprinklers. The average rate at which water is sprayed onto the crops is measured in mm/hour.

The application rate depends on the size of sprinkler nozzles, operating pressure and distance between the sprinklers. The application rate must not exceed the maximum allowable infiltration rate for the soil type. Excess application rate will result in water loss, soil erosion and surface sealing. There may be inadequate moisture in the root zone of crops or plants after irrigation and they may get damaged.

The force with which the water flows out of the sprinkler is known as its 'water pressure'. Water pressure is measured in pounds per square inch (psi). Sprinklers are, therefore, designed to work at certain pressure levels, which are recommended as their operating pressure. If the pressure is above or below than the recommended level, then the distribution of water will be affected. When the pressure is low, the water drops become larger and they cannot irrigate the crops that are far from the system. If the pressure is high, then the droplets will be smaller and the crops will not be irrigated evenly. It can also damage the sprinkler heads. Although sprinklers are adaptable to most soils, they are best suited for sandy soil. These can be used for irrigating lawns, gardens and agricultural fields.

Types of sprinkler irrigation system

(i) Centre pivot

(ii) Towable pivot

(iii) Rain gun

(iv) Impact sprinkler

(v) Pop up sprinkler

(vi) Linear move sprinkler

Centre pivot

The centre pivot is capable of irrigating most field crops. It consists of a single sprinkler lateral supported by a series of towers. It is anchored at one end and rotates around a fixed central point called 'pivot point'. The control panel attached to the pivot point gives commands to the central pivot machine. A drive unit or drive tower touches the ground, which contains necessary components for the machine to move. It consists of a base beam, drive train, wheels and other structural support equipment. The towers are self propelled so that the lateral rotates around the pivot point installed in the centre of the irrigated area. The long pipes between the drive units are called 'spans'. Spans consist of the main water line, sprinklers and a supporting structure to hold the weight between the towers. A tower box controls the drive unit components, with regard to the direction and duration.

Towable pivot

Towable pivot is similar to centre pivot. But here, the pivot is towed away by a tractor. There are 3–4 wheels in the centre of the pivot, which make it possible to move the pivot from one place to another by pulling it with the help of the tractor. It helps farmers to carry out mechanised irrigation in an economical manner. It can easily irrigate fields as the machine can be towed away from one field to another in minimum time.

Rain gun

A rain gun is used as a water spray mist or fog beam. It discharges water at less than 175 lph. It is used to irrigate trees and other crops separated widely. Fruit tree crops like citric fruits, mango, guava, avocado, etc., can be irrigated with a rain gun. The passage diameter of the rain gun is small. Therefore, the release of filtered water is essential, amounting up to a requirement of 60–80 mesh. The minimum operating pressure is 1.5–2 kg/cm^2. The heads of rain gun are mounted on plastic wedges/piles 20–30 cm above the ground. Rain gun is suitable for field crops like groundnut, onion, potato, sugarcane, cotton and plantation crops, such as coffee and tea.

Impact sprinkler

This sprinkler is driven in a circular motion by the force of outgoing water, and at least, one of its arms extends from the head. The sprinkler arm is repeatedly pushed back into the water stream by a spring. When the arm strikes the water stream, it scatters the stream and re-orients the flow, enabling a uniform watering area around the sprinkler. Impact sprinkler is recommended for closely spaced field crops like potato, leafy vegetables, cotton, oilseeds, pulses, cereals, fodder crops, etc.

Pop up sprinkler

A pop up sprinkler consists of an inlet, body, cap, wiper seal, riser, nozzle and radius adjustment screw. Such a sprinkler is portable and easy to install, thereby, making it ideal for irrigating lawns, seasonal flowers and planting beds.

Linear move sprinkler

Linear move sprinkler irrigation system is similar to the centre pivot system in construction, except that neither end of the lateral pipeline is fixed. It is composed of a series of towers that are suspended and move laterally in the direction of rows. The whole line moves down the field perpendicular to the lateral. Water delivery to the continuously moving lateral is by a flexible hose or open ditch pickup. Both the centre pivot and linear move systems are capable of high efficiency water application. By 'water efficiency', it means reducing water wastage by measuring the amount of water required for a specific purpose and the amount of water delivered or used. Such a system requires high capital investment but is not labour intensive.

3. Micro-sprinkler irrigation system

Introduction

Micro-sprinklers provide irrigation using very fine droplets. They have rotating deflector, called rotor or dancer, who helps in delivering greater diameter coverage, lower precipitation rate than diffusers, increased droplet size, and better distribution of water. For each type of micro-sprinkler are several types of rotors.

Micro-sprinkler irrigation has become established and widely used in recent years due to its potential to increase yields and due to increased irrigation efficiency and decreased labor requirements.

Micro-sprinkler irrigation applies water directly to the soil surface area allowing water to dissipate under low pressure and infiltrate the soils in wetted profile that uniformly meets water demand throughout.

This irrigation system can be configured in one of two ways with:

- ❖ drop down sprinklers from suspended irrigation lines
- ❖ risers mounted on stakes from surface irrigation lines

Both above methods connect sprinklers to in-row irrigation lines with spaghetti tubing. Drop down sprinklers allow for mowing between trees since no moving of sprinklers is required to mow, unlike with riser mounted sprinklers. Such irrigation systems are useful and suitable for sloping or irregularly-shaped pieces of land that are otherwise impractical to furrow irrigate.

Micro-sprinkler systems usually have sprinkler between every other tree and are staggered in adjoining tree rows to maximize irrigation coverage and efficiency, and minimize costs. The size of the sprinkler emitter, or orifice and working pressure, determines application rate. Maximum application rates are in turn dependent on soil characteristics such soil texture and maximum water infiltration rates for given soils, so as not to apply water at a faster rate than the soil can absorb in a given time. When irrigation water contains sediment, typically ditch or reservoir water, micro-sprinkler systems usually make use of an in-line filter system to prevent emitter clogging and reduce wear.

Micro-irrigation systems can apply water on short-set, high-frequency basis, optimal for younger trees with smaller root mass, or long-set, lower frequency irrigations to provide more consistent and optimal soil moisture environment for mature trees, as per orchard and water supply situation. Soil moisture monitoring devices, such as composite gypsum blocks or Watermark sensors, helps in improving overall irrigation system performance by allowing irrigator to monitor and manage soil moisture more conveniently and accurately.

Since micro-sprinkler irrigation systems apply water in a way that can be very precise at meeting crops water needs, it is recommended that irrigators practice the 'water balance approach' regardless of irrigation frequency. Such approach involves calculating the daily water use by crop and replenishing it on an as needed basis depending on soil water holding capacity and the orchard's age, top soil depth, rooting depth, and water availability schedule.

System Layout and Equipment

Micro-irrigation systems consist of system head and distribution network which includes:

- pump
- filter
- flow meter
- pressure gauges
- fertilizer injector
- pressure regulator
- controller

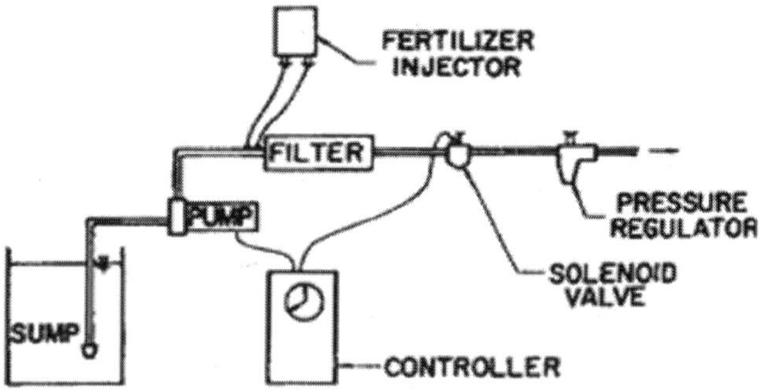

The meter and acid injector are optional equipment but highly desirable as they help in monitoring of system performance and add flexibility to system. The distribution network consists of pipes made of polyethylene, pipe fittings, sprinklers and valves. Valves can be actuated electrically by a controller connected to a solenoid valve in the case of an automated system or manually.

When sediment is found in irrigation water, water filtration is essential for protecting sprinkler nozzles from clogging and the irrigation system from rapid wear. Two basic filters like sand media filters and screen/spin filters are required for one stage of filtration, though micro irrigation sprinkler systems usually require less filtration than drip irrigation systems, filtration is a must if not using well water. The bigger the orifice of the sprinkler, the less filtration is needed.

Micro-sprinkler irrigation systems operate at relatively low pressure compared to large sprinkler irrigation systems. Due to this, pumping costs are substantially less. A pressure regulator is used to control the line pressure as sprinklers have a maximum operating pressure for optimal efficiency. Many pressure regulators are desirable for locations with large elevation changes, as pressure increases with elevation drops. Small diameter polyethylene pipe (1/2 to 1 inch) is generally used for the in-row laterals that are laid on the soil surface with risers or suspended on wire and T-posts in the tree row with drop down sprinklers nozzles, typically sprinklers are connected to the poly pipe with spaghetti tubing. Irrigation lines are buried between rows to facilitate tractor operations. The lateral is connected to a manifold that is supplied with water through a main and/or sub-main connection. Manifolds, sub-mains and mains are usually buried with control valves either above or below ground.

The purpose of the lateral is to supply water to sprinklers located in the tree row and the lateral should be sized for the maximum flow rate of water application. Sprinklers are chosen for flow rate at the tree but are dependent on lateral pressure to perform correctly. Sprinkler manufacturers specify minimum and maximum pressure ranges to maximize system efficiency for each sprinkler. Once sprinklers are selected and installed, lateral pressure is controlled by a pressure regulator valve to match sprinkler size and capacity. It is important that all sprinklers being fed from a common lateral are selected for the same pressure. Sprinkler dealers can size and design complete sprinkler irrigation systems for the exact needs of your orchard and your water supply.

Features:

- Micro-sprinkler irrigation is a low pressure, low to medium volume irrigation system suitable for high value crops such as tree fruits.
- If managed properly, such irrigation system can increase yields and decrease water use and fertilizer and labor requirements when compared to gated pipe/furrow irrigation systems.
- Micro-sprinkler irrigation saves water due to high application efficiency and high water distribution uniformity with little if any waste if managed properly.
- Micro-sprinkler irrigation is ideal for irrigating sloping or irregularly shaped orchard blocks that cannot be flood/furrow irrigated.
- In this, there will be extra increase in efficiency, when fertilizer is injected to the system. Water-soluble fertilizers can be injected through a micro irrigation system thereby significantly reducing cost compared to traditional fertilizer applications.

Advantages

- ❖ It has a potential for reducing frost damage
- ❖ It has greater control over water application times and amounts to more closely match local evapotranspiration rates to maintain optimum soil moisture.
- ❖ It has lower susceptibility to clogging due to large orifice sizes as coompared to drip irrigation system.
- ❖ It reduces the need for extra fine filtration.

Applications

- ❖ Such type of sprinkler system are ideal for low volume irrigation in horticultural crops, fruit, flowers, greenhouses, nurseries, frost protection and watering gardens.
- ❖ Such system allow application of pesticides in plant cover crops.
- ❖ Its use is widespread in greenhouses, especially in leaf vegetables.

4. Mini-sprinkler irrigation system

Introduction

Mini Sprinkler system is designed for wide range of general field uses which uses portable, semi-portable and solid set systems, in order to meet the demand of modern, economic irrigation systems. It has sophisticated field tested construction and uses highly corrosion resistant materials, ensuring safe operation and long life with less maintenance.

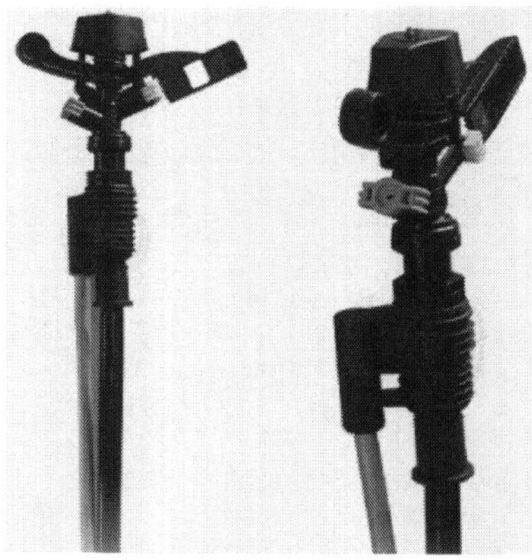

Components

The major parts of mini sprinkler irrigation system are:

- ❖ Sprinkler Nozzle
- ❖ Pump
- ❖ Foot batten
- ❖ Riser pipe

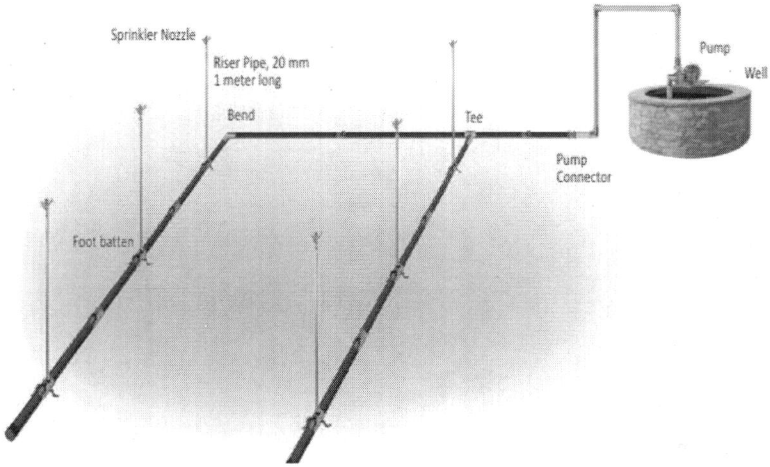

Advantages

- ❖ Precise application and adjustment of quantity and rate possible.
- ❖ Low pumping costs, operating at the same pressure as drip irrigation.
- ❖ Larger wetted zone than with drip irrigation thus plants are less likely to suffer from water stress should there be any delay in irrigation.
- ❖ Leaf disease impact is uncertain, leaves remain dry but relative humidities will be enhanced.
- ❖ Fertigation and chemigation can be utilized.
- ❖ Intermediate wetted area regarding weed control compared to drip and overhead sprinklers.
- ❖ Other management tasks generally not obstructed during irrigation application.
- ❖ Suitable for automation.
- ❖ Efficient utilization of limited water supplies and no runoff.
- ❖ Filtration is required but is not as critical as for drip irrigation.

1.9 Reading signs, notices and/or cautions at site

Micro-irrigation technician should see and follow local safety sign and regulations at site which are applied for installing, operating, maintaining and troubleshooting of drip irrigation system and its components. He/she should regularly follow the safety signs which are marked at the irrigation sites such as:

WARNING 

This warning sign gives an indication that worker while working in agricultural environment should always wear protective footwear.

WARNING 

This warning sign gives an indication that only authorized electricians are allowed to do the electrical installations which should be done with local safety standards and regulations.

WARNING 

This warning sign gives an indication that worker should take measures to prevent the infiltration of nutrients, acids and chemicals in water source.

ACID HAZARD 

This warning sign gives an indication that:

If nutrients, acids and chemicals are not handled properly, it cause serious injury/death and may also damage the crop, soil, environment and irrigation system, so proper handling of nutrients, acids and chemicals is the responsibility of the grower/technician.

Always observe nutrient, acid, chemical manufacturer's instructions and regulations issued by the relevant local authority.

WARNING  

This warning sign gives an indication that worker while handling nutrients, parts, acids and chemicals should always use protective equipment like gloves and goggles to safeguard themselves.

CAUTION 

This warning sign gives an indication to technician/worker that while opening or closing any manual valve, always do it gradually in order to prevent damage to the system by water hammer.

Chapter 2

Sprinkler Irrigation System

2.1 Introduction to sprinkler irrigation

2.1.1 Introduction

Sprinkler irrigation is a method of applying irrigation water which is similar to natural rainfall. In this, water is distributed through a system of pipes by pumping which is sprayed into air by sprinklers so that it breaks up into small water drops that fall to the ground. In this, the pump supply system, sprinklers and operating conditions must be designed to enable uniform application of water.

Such type of irrigation system is to be used when:

Crops are suitable:

- ❖ Sprinkler irrigation is suited for most row, field and tree crops and water can be sprayed over or under the crop canopy.
- ❖ Large sprinklers are not recommended for irrigation of delicate crops as large water drops produced by the sprinklers may damage the crop.

Slopes are suitable:

- ❖ Sprinkler irrigation is adaptable to any farmable slope, whether uniform or undulating.
- ❖ The lateral pipes supplying water to the sprinklers should always be laid out along the land contour whenever possible which minimizes pressure changes at sprinklers and gives uniform irrigation.

Soil is suitable:

- ❖ Sprinklers are best suited to sandy soils with high infiltration rates although they are adaptable to most soils.
- ❖ The average application rate from the sprinklers is always chosen to be less than the basic infiltration rate of the soil so that surface ponding and runoff can be avoided.
- ❖ Sprinklers are not suitable for soils which easily form a crust.
- ❖ If sprinkler irrigation is only method available, then light fine sprays should be used.

The larger sprinklers produces larger water droplets are to be avoided.

Irrigation water is suitable:

❖ A good clean supply of water, free of suspended sediments, is required to avoid problems of sprinkler nozzle blockage and spoiling the crop by coating it with sediment.

2.1.2 Objectives of Sprinkler Irrigation

The main objective of a sprinkler system is to apply water as uniformly as possible to fill the root zone of the crop with water.

Wetting Patterns

The wetting pattern from a single rotary sprinkler is not very uniform. Normally the area wetted is circular.

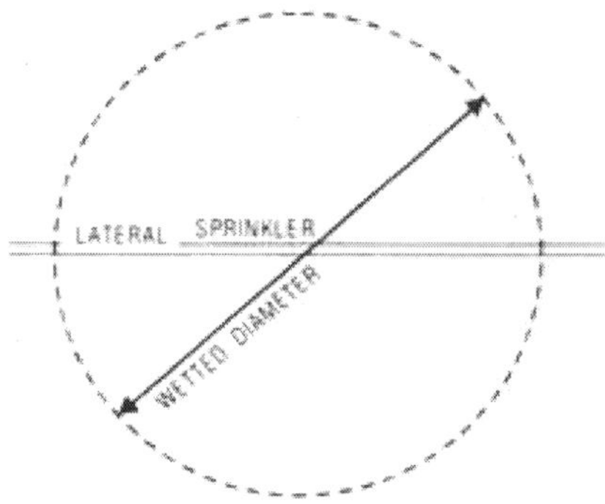

The heaviest wetting is close to the sprinkler. For good uniformity several sprinklers must be operated close together so that their patterns overlap.

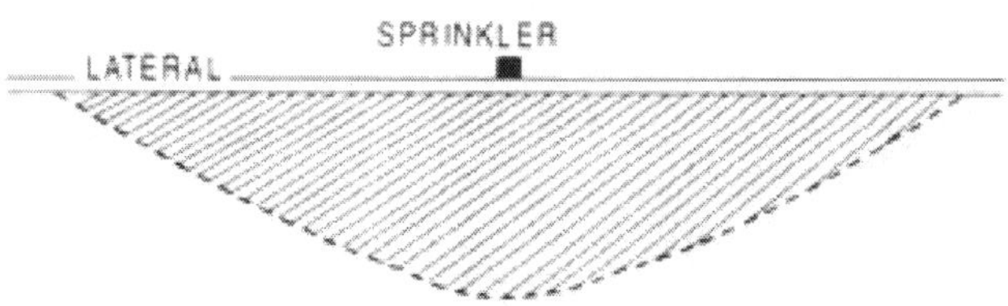

For good uniformity the overlap should be at least 65 per cent of the wetted diameter. This determines the maximum spacing between sprinklers.

Wetting Patterns for Single Sprinkler

The uniformity of sprinkler applications can be affected by wind and water pressure. Spray from sprinklers is easily blown out by even a gentle breeze and this can seriously reduce uniformity. To reduce the effects of wind the sprinklers can be positioned more closely together.

Sprinklers will only work well at the right operating pressure recommended by the manufacturer. If the pressure is above or below this then the distribution will be affected.

The most common problem is when the pressure is too low. This happens when pumps and pipes wear. Friction increases and so pressure at the sprinkler reduces.

The result is that the water jet does not break up and all the water tends to fall in one area towards the outside of the wetted circle. If the pressure is too high then the distribution will also be poor. A fine spray develops which falls close to the sprinkler.

Application Rate

This is the average rate at which water is sprayed onto the crops (mm h-1). The application rate depends on the size of sprinkler nozzles, the operating pressure and the distance between sprinklers. When selecting a sprinkler system it is important to make sure that the average application rate is less than the basic infiltration rate of the soil. In this way all the water applied will be readily absorbed by the soil and there should be no runoff.

Sprinkler Drop Sizes

As water sprays from a sprinkler it breaks up into small drops between 0.5 and 4.0 mm in size. The small drops fall close to the sprinkler whereas the larger ones fall close to the edge of the wetted circle. Large drops can damage delicate crops and soils and so in such conditions it is best to use the smaller sprinklers.

Drop size is also controlled by pressure and nozzle size. When the pressure is low, drops tend to be much larger as the water jet does not break up easily. To avoid crop and soil damage use small diameter nozzles operating at or above the normal recommended operating pressure.

2.1.3 System Layout

The sprinkler irrigation system includes measurement of land and drawing its sketch to scale on a graph paper where location of hedges, shrubs, trees, walls and driveways are marked. In the sketch, the area is divided into zones for laying out main lines and laterals. Now after this, amount of available water flow for sprinkler system is determined so as to ascertain how many sprinkler heads can run at a time. In this, pipe layout is to be planned as per the placement of risers and sprinkler heads. In sprinkler irrigation, water is conveyed under pressure through pipes to the area to be irrigated, where it is discharged through sprinklers.

Sprinkler irrigation system consists of:

- pump unit
- main line
- laterals
- risers
- sprinkler heads with filter screens
- desilting devices
- flow regulators
- fertilizer application system

Pump unit

- In this, a centrifugal pump takes water from the source and provides adequate pressure for delivery in pipe system.
- A pump is used for developing the required pressure which can be used when land is undulating for leveling, soil is porous, erodible and impermeable and flow rate is too less for employing surface irrigation method.

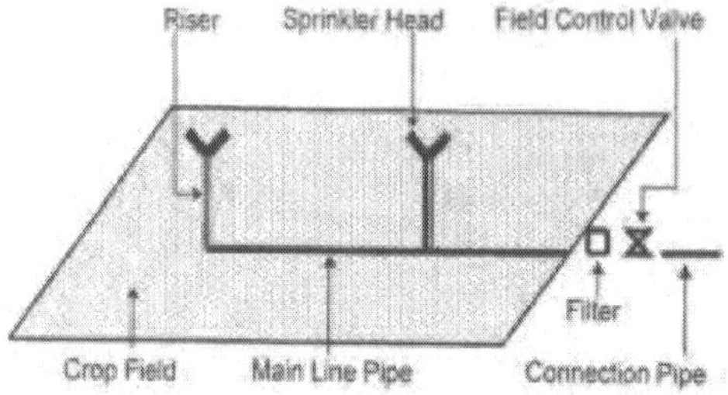

Mainline and sometimes submainlines

- The mainline/sub mainlines are pipes that deliver water from the pump to the laterals.
- The layout of mains, sub-mains and laterals depends on local conditions like topography, soil characteristics and source of water.
- The main line is laid along the slope and the laterals across the slope or nearly on the contours.
- These pipelines are permanent and are laid on the soil surface or buried below ground.
- Such pipelines are temporary, and can be moved from field to field.
- The main pipe materials used include asbestos cement, plastic or aluminium alloy.

Laterals

- The laterals deliver water from mainlines/submainlines to the sprinklers.
- They can be permanent but more often they are portable and made of aluminium alloy or plastic so that they can be moved easily.
- In portable system, laterals need to be of the same size so that they can be changed easily.

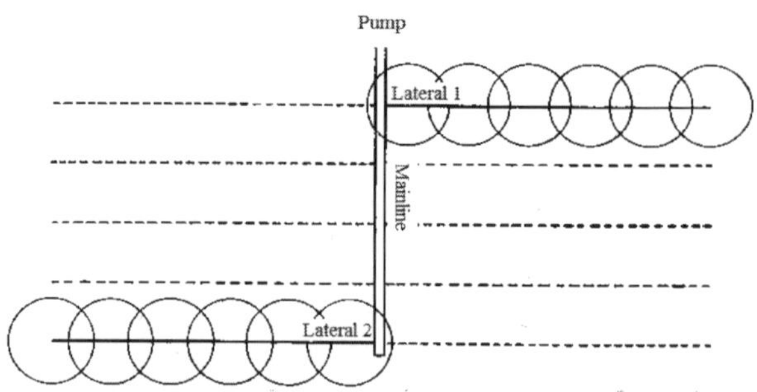

- In this, lightweight aluminium or plastic pipes are moved by hand.
- In this, rotary sprinklers are spaced 9-24 m apart along the lateral which is normally 5-12.5 cm in diameter.
- The lateral pipe is located in the field until the irrigation is complete.
- The pump is switched off and lateral is disconnected from the mainline and moved to the next location where it gets re-assembled and connected to mainline which starts irrigation again.
- The lateral can be moved 1-4 times in a day and can be gradually moved around the field till the whole field is irrigated.

Sprinklers

- The type of sprinkler depends on its nozzle size and pressure with which it discharges water.
- It must also be ensured that the water discharged does not cause run-off or damage to the crops.
- Sprinkler should supply water to the crops sown in a field uniformly under the prevailing wind conditions.
- It must meet the irrigation water requirement of a crop and irrigation frequency.
- The common symbols used when designing sprinkler irrigation plans can be:
    - Sprinkler – full: It is a nozzle which throws water all around it at 360° and at a distance of 3.6–4.5 metre.
    - Sprinkler – half: It is a nozzle which throws water all around it at 180° and at a distance of 3.6–4.5 metre.
    - Sprinkler – quarter: It is a nozzle which throws water all around it at 90° and at a distance of 3.6–4.5 metre.

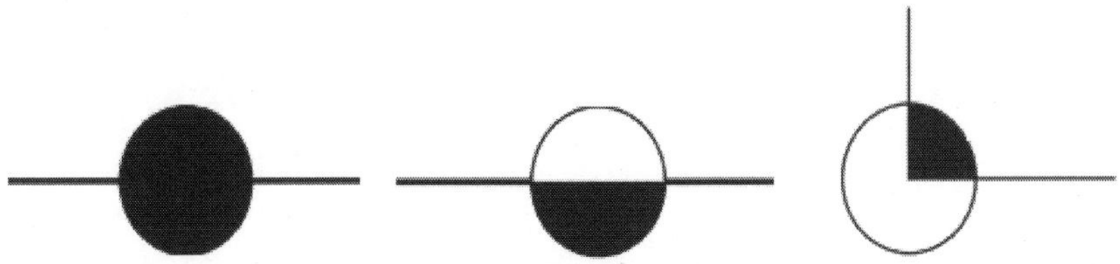

sprinkler – full *sprinkler – half* *sprinkler – quarter*

- Sprinkler – one-third: It is a nozzle which throws water all around it at 120° and at a distance of 3.6–4.5 metre.
- Sprinkler — three quarter: It is a nozzle which throws water all around it at 270° and at a distance of 3.6–4.5 metre.
- Sprinkler — two-third: It is a nozzle which throws water all around it at 240° and at a distance of 3.6 to 4.5 metre.

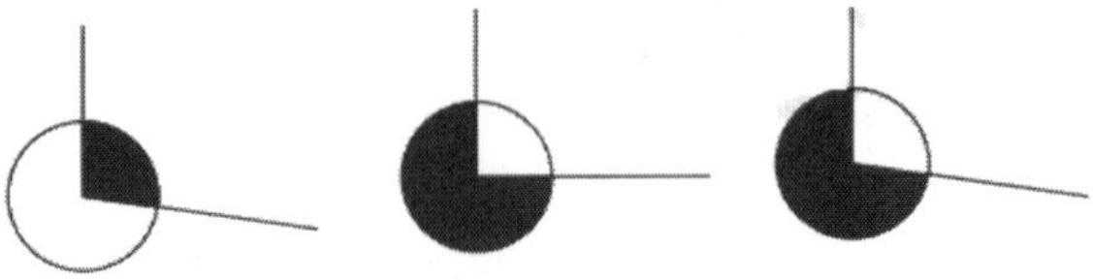

sprinkler – one-third sprinkler – three quarter sprinkler – two-third

2.1.4 Principles of Sprinkler Irrigation Systems:

It is a method of applying irrigation water which is similar to natural rainfall. Sprinkler irrigation water is applied through a pressurized system. The pressure causes the water to flow out through the sprinkler nozzle and fly through the air so that it breaks up the water into small water drops which falls on the soil surface.

The main objective of a sprinkler system is to apply water as uniformly as possible to fill the root zone of the crop with water. This system is suited for most row, field and tree crops. However, large sprinklers are not recommended for irrigation of delicate crops such as lettuce because the large water drops produced by the sprinklers may damage the crop. Sprinklers are best suited to sandy soils with high infiltration rates although they are adaptable to most soils.

If sprinkler irrigation is the only method available, then light fine sprays should be used. The larger sprinklers producing larger water droplets are to be avoided. A good clean supply of water, free of suspended sediments, is required to avoid problems of sprinkler nozzle blockage and spoiling the crop by coating it with sediment.

2.2 Benefits of sprinkler irrigation

When the irrigation water is applied to the land in form of spray, through a network of pipes and pumps, then such method is known as sprinkler irrigation method. Such type of irrigation method can be used for all crops and all types of soils.

There are certain benefits of using sprinkler irrigation method which includes:

- The method is useful for large range of topographic conditions, soils and crops.
- In this equal water is given to all crops.
- In this, borders, field channels, etc. does not require much area, so lot of area is available for cropping.
- With this system, soil erosion can be controlled.
- In this, labor cost is much less as there is no need for land preparation.
- Such irrigation method is beneficial for seedling and plants which are very young.
- In this, time and amount of use of fertilizer can be better controlled.
- Such type of irrigation system is especially adaptable to more humid regions.
- With such system, high and frequent application can be effectively accomplished.
- It has easy mechanization and automation.

2.3 Limitations of sprinkler irrigation

There are certain limitations of using sprinkler irrigation method which includes:

- It has high operating cost.
- In this, water will drift when there is lot of wind.
- It needs stable water supply.
- In this, saline water can burn the leaves of crops which may cause problem.
- In this, water must be free from sand, debris and large amount of salt.
- In such system, sensitivity to wind causes evaporation losses.
- Such type of irrigation system is costly to install.
- The system has high maintenance cost.
- With underground pipes, automatic sprinkler head function, electronic components, timers, controllers and other parts, can go wrong.
- In this, the performance gets affected by temperature.
- In this, metal contaminated wastewater can't be treated.

2.4 Components of sprinkler irrigation system

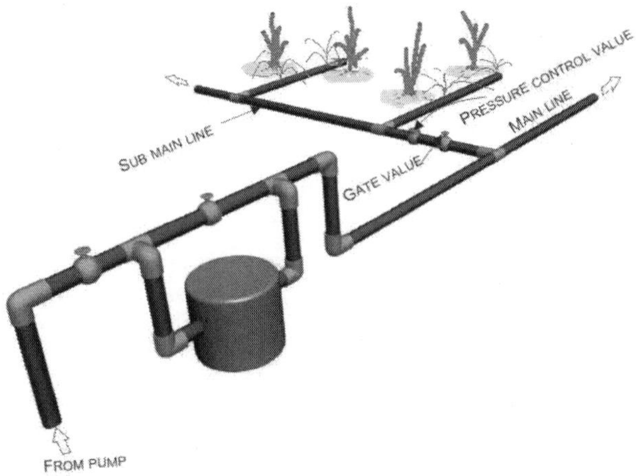

A typical sprinkler irrigation system comprises of:

- Pump unit
- Mainline and sometimes sub-main lines
- Laterals
- Couplers
- Sprinkler Head
- Fittings and accessories- valves, bends, plugs and risers

Pump Unit:

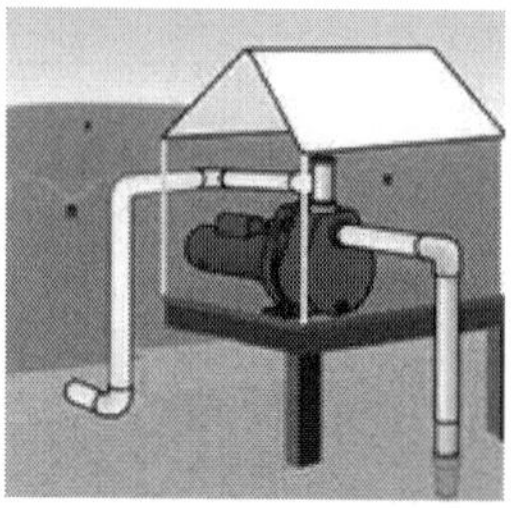

- It is usually a centrifugal pump, which takes water from the source and provides adequate pressure for delivery into the pipe system.
- In this, water from the source is pumped under pressure to sprinkler system which is created by pump forces water through sprinklers or by perforations/nozzles in pipelines which forms a spray.
- A high speed centrifugal or turbine pump can be used for operating sprinkler irrigation to individual fields.
- Centrifugal pump is used when distance from the pump inlet to the water surface is less than 10 meters.
- For pumping water from deep wells or more than eight meters, a submersible pump is used.
- The driving unit of pump may be either an electric motor or an internal combustion engine.

Mainline and Sub-Mainlines:

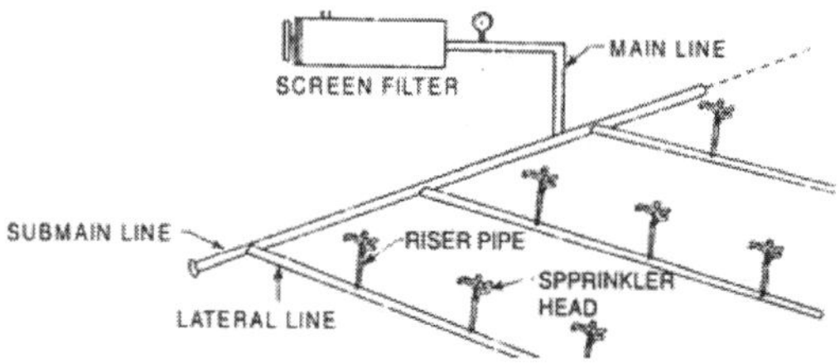

- These are pipes, which deliver water from the pump to the laterals.
- The pipelines are permanent and are laid on the soil surface or buried below ground.
- These are temporary, and can be moved from field to field.
- The main pipe materials used include plastic or aluminium alloy.

Laterals:

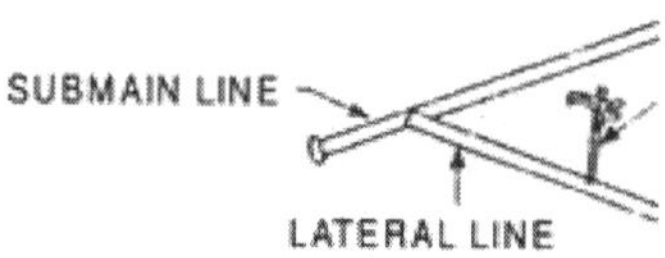

- ❖ The lateral deliver water from the mainlines or sub-mainlines to the sprinklers.
- ❖ Laterals can be permanent, but more often, they are portable and made of aluminium alloy or plastic so that they can be moved easily.

Couplers:

- ❖ They are used for connecting two pipes and uncoupling quickly and easily.
- ❖ It should provide reuse and flexible connection with leak proof joint.
- ❖ It should be simple and easy to couple and uncouple pipes.
- ❖ Coupler should be light, non-corrosive and durable.

Sprinkler Head:

- ❖ It distributes water uniformly over the field without runoff or excessive loss due to deep percolation.
- ❖ It is most important components of sprinkler irrigation system where its operating characteristics under optimum water pressure and climatic conditions determines the suitability and efficiency of the system.
- ❖ There are different types of sprinklers available which can be rotating or fixed type which depends on actual irrigation purpose and plot sizes.
- ❖ The rotating type sprinkler can be adapted for wide range of application rates and spacing which are effective with pressure of 10-70 m head at sprinkler.
- ❖ Fixed head sprinklers are commonly used to irrigate small lawns and gardens.

Fittings and Accessories:

The following are important fittings and accessories used in sprinkler system:

- ❖ Water Meter: It is used to measure the volume of water delivered. The meter is necessary to operate the system to give the required quantity of water.
- ❖ Flange, couplings and nipple used for proper connection to the pump, suction and delivery.
- ❖ Pressure Gauge: It gives information about working of sprinkler system with desired pressure to ensure application uniformity.
- ❖ Bend, tees, reducers, elbows, hydrants, butterfly valve and plugs.
- ❖ Fertilizer Applicator: Soluble chemical fertilizers can be injected in sprinkler system which is applied to the crop. The equipment for fertilizer application is cheap and simple and can be fabricated locally. The fertilizer applicator has sealed fertilizer tank with tubings and connections.

2.5 Types of sprinkler irrigation systems

There are many types of sprinkler systems and there are several versions of each type. The main types are periodic move systems which can be hand-move, end-tow, and side-roll laterals; side-move laterals with or without trail lines; and gun and boom sprinklers.

The fixed systems typically use either small or gun sprinklers. The main type is continuous-move systems which can be center-pivots, traveling gun or boom sprinklers, and linear move or lateral move.

Micro spray systems are also sprinkler systems whose hydraulics, design and operations are similar to those of drip or trickle systems.

2.5.1 Periodic move sprinkler systems

(1) Hand-move lateral

Hand-move laterals are composed of separate lengths of aluminum tubing with quick couplers that have either center-mounted or end-mounted riser pipes with sprinkler heads. Hand-move laterals are often referred to as hand lines. Water is conveyed to them with portable or buried mainline pipe with valve outlets at constant spacing. Systems are composed of one or more laterals. In previous years, this type of system was used to irrigate more area than any other system, and is still used on almost all crops and types of topography. Major disadvantages of the system are its high labor and high pressure requirements. This system is the basis from which all of the mechanized systems were developed.

Typical lengths of hand-move laterals are 1,280 feet and typical lateral spacing, the distance the lateral is moved each time, are 50 and 60 feet. Spacing of sprinklers along laterals is generally 30 to 40 feet, which normally corresponds to the separate lengths of pipe. Laterals are generally constructed of aluminum for ease of movement. Some lateral systems use polyethylene pipe. Mainline systems are often buried, with risers extending to the ground surface each 50 to 200 feet to which laterals are connected. Typical sprinkler operating pressures are 40 to 60 pounds per square inch. Crops irrigated using hand-move systems may need to have sufficiently deep roots or high-soil water-holding capacity, such that the laterals do not have to be returned to a specific location within less than about 5 days, but this depends on the availability of sprinkler pipe, labor, and water supply flow rate.

The job of moving a hand-move system requires more than twice the amount of time per irrigated acre and is not nearly as easy as the job of moving an end-tow, side-roll, or side-move system. If the wet soil is soft or sticky, even more labor is required to move each section of a hand-move system. Nevertheless, a major inconvenience of mechanical move systems occurs when the laterals reach the end of an irrigation cycle.

When this happens with a hand-move system, the laterals at the field boundaries can be disassembled, loaded on a trailer, and hauled to the starting position at the opposite boundary. A mechanical-move lateral cannot be disassembled so easily, so an irrigator might decide to deadhead it back to its starting position. This operation is time consuming, especially when trail tubes are involved. However, in practice it is not uncommon to disassemble side-roll laterals after each irrigation, especially when labor costs are not prohibitive.

(2) End-tow lateral

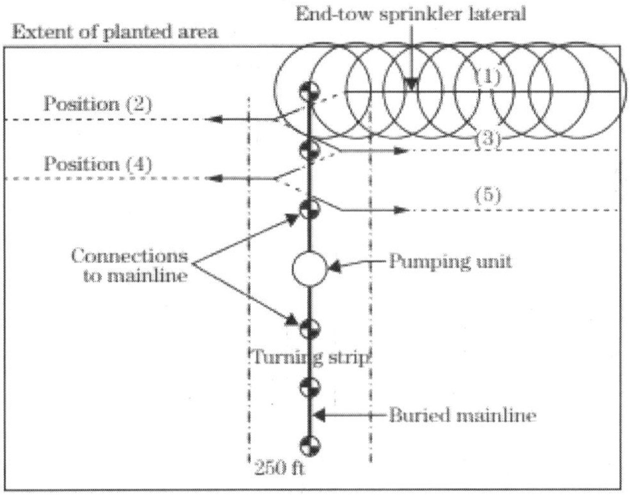

An end-tow lateral system is similar to one with hand move laterals except the system consists of rigidly coupled lateral pipe connected to a mainline. The mainline should be buried and positioned in the center of the field for convenient operation. Laterals are towed lengthwise over the mainline from one side to the other. After draining the pipe through automatic quick drain valves, a 20- to 30-horsepower tractor can easily pull a quarter-mile, 4-inch (100 mm) diameter lateral.

Two carriage types are available for end-tow systems. One is a skid plate attached to each coupler to slightly raise the pipe off the soil, protect the quick drain valve, and provide a wear surface when towing the pipe. Two or three outriggers are required on a quarter mile lateral to keep the sprinklers upright. The other type uses small metal wheels at or midway between each coupler to allow easy towing on sandy soils.

End-tow laterals are the least expensive mechanical move systems; however, they are not well adapted to small or irregular areas, steep or rough topography, row crops planted on the contour, or fields with physical obstructions. They work well in grasses, legumes, and other close-growing crops and fairly well in row crops, but the laterals can be easily damaged by careless operation such as moving them before they-have drained, making too sharp an "S" turn, or moving them too fast. They are not, therefore, recommended for projects where the quality of the labor is undependable.

When used in row crops, a 200- to 250-foot-wide turning strip is required along the length of the mainline. The turning strip can be planted in alfalfa or grass. Crop damage in the turning areas can be minimized by making an offset equal to one-half the distance between lateral positions each time the lateral is towed across the mainline instead of a full offset every other time. Irrigating a tall crop such as corn requires a special crop planting arrangement such as 16 rows of corn followed by four rows of a low growing crop that the tractor can drive over without causing much damage.

A relatively new type of end-tow sprinkler system uses sprinklers in pods, spaced evenly along a line with skid plates which help hold the pods in an upright position. The line is pulled by a tractor or other vehicle to each irrigating position, but it is not moved while in operation. The pods can be designed to prevent damage due to livestock or horses stepping on the sprinklers, and they are typically used in pasture irrigation.

(3) Side-roll lateral

A side-roll lateral system, often referred to as wheel lines, is similar to a system with hand-move laterals. The lateral pipes are rigidly coupled together, and each pipe section is supported by a large wheel. The lateral line forms the axle for the wheels, and when a torque is applied the line rolls sideways. This unit is moved mechanically by an engine mounted at the center of the line or by an outside power source at one end of the line.

Side-roll laterals work well in low-growing crops. They are best adapted to rectangular fields with fairly uniform topography and no physical obstructions, but are often used on irregular field shapes. The diameter of the wheels should be selected so that the lateral pipe clears the mature crop and so that the specified lateral move distance is a whole number of rotations of the line. For example, for a 60 foot move, use three rotations of a 76.4 inch diameter wheel, and for a 50 foot move, use three rotations of a 64-inch diameter wheel. In some cases a 6-foot-wheel diameter is used to provide sufficient crop clearance, whereby one rotation of the wheel is approximately 20 feet, so that five revolutions is approximately 100 feet. Then, in one irrigation the sets are in alternating moves of two and three revolutions, and in the next irrigation they are alternating sets of three and two revolutions, thereby providing an effective lateral spacing of about 50 feet for each pair of irrigations.

Side-roll laterals up to 2,000 feet or longer can be satisfactory for use on close-planted crops and smooth topography. For rough or steep topography and for row crops with deep furrows, such as potatoes, laterals up to a quarter-mile long are recommended, but the maximum length depends on many factors, such as the crop density, field topography, and pipe strength.

Spacing of sprinklers along laterals is generally 30-40 feet. Typically, 4-5 inch diameter aluminum tubing is used for laterals. For a standard quarter mile lateral on a close-spaced crop at least three lengths of pipe to either side of a center power unit should be 0.072 inch heavy walled aluminum tubing to prevent the pipe from twisting and collapsing during changes in set. For longer lines and in deep-furrowed row crops or on steep topography heavier walled tubing should be used, enabling the laterals to roll more smoothly and uniformly, and with less chance of breaking. Mainline systems and operating pressures are as previously described for hand lines.

A well designed side-roll lateral should have quick drains at each coupler so that it can be turned off, drained, and moved within 30 minutes. All sprinklers should be provided with a self leveler so that each sprinkler will be upright, regardless of the position at which the lateral pipe is stopped. In addition, the lateral should be provided with at least two wind braces, one on either side of the power mover and with a flexible or telescoping section to connect the lateral to the mainline hydrant valves.

Trail tubes, or drag lines, are sometimes added to heavy walled 5-inch side-roll laterals. With sprinklers mounted along the trail tubes, the system has the capacity to irrigate more land than the conventional side-roll laterals. Special couplers with a rotating section are needed so the lateral can be rolled forward. Quick couplers are also required at the end of each trail tube so they can be detached when a lateral reaches its last operating position. The lateral must be rolled back to the starting location where the trail tubes are, then reattached for the beginning of a new irrigation cycle.

(4) Side-move lateral

Side-move laterals, relatively rare, are moved periodically across the field in a manner similar to side-roll laterals. An important difference is that the pipeline is carried above the wheels on small A-frames instead of serving as the axle. Typically, the pipe is carried 5 feet above the ground and the wheel carriages are spaced 50 feet apart. A trail tube with 11 sprinklers mounted at 30 foot intervals is pulled behind each wheel carriage. Thus, the system can wet a strip 320 foot wide, allowing a quarter-mile long line to irrigate approximately 11 acres at a setting. This system produces high uniformity and low application rates.

Side-move lateral systems are suitable for most field and vegetable crops. For field corn, however, the trail tubes cannot be used, and the A-frames must be extended to provide a minimum ground clearance of 7 feet. Small gun sprinklers mounted at every other carriage will irrigate a 150-foot wide strip, and a quarter-mile-long lateral can irrigate 4.5 acres per setting. Application rates, however, are relatively high.

The three main types of fixed systems are those with solid-set portable hand-move laterals, buried or permanent laterals, and sequencing valve laterals. Most fixed-sprinkler systems have small sprinklers spaced 30 to 80 feet apart, but some systems use small gun sprinklers spaced 100 to 160 feet apart.

Solid-set portable

Fig Hand-move sprinkler laterals in a solid-set

Solid-set portable systems are often used for potatoes and other high-value crops where the system can be moved from field to field as the crop rotation or irrigation plan for the farm is changed. These systems are sometimes moved from field to field to germinate such crops as lettuce that require irrigation up to three times per day, but that are then furrow irrigated following germination and root establishment. Moving the laterals into and out of a field requires much labor, although this requirement can be reduced by the use of special trailers on which the portable lateral pipe can be stacked by hand. After a trailer has been properly loaded, the pipe is banded in several places to form a bundle that is lifted off the trailer at the farm storage yard with a mechanical lifter. The procedure is reversed when returning the laterals to the field for the next season. In some cases where fields are flat and furrows are large, the laterals can be floated out of the germinated field using furrow irrigation. The laterals are disassembled section by section at the field edge. This reduces the need for foot traffic in wet fields and reduces labor requirements.

Buried laterals

Permanent, buried laterals are placed underground 18 to 30 inches deep with only the riser pipe and sprinkler head above the surface. Many systems of this type are used in citrus groves, orchards, vineyards, and other crops of relatively high value.

The sequencing valve lateral may be buried, laid on the soil surface, or suspended on cables above the crop. The heart of the system is a valve on each sprinkler riser that turns the sprinkler on or off when a control signal is applied. Some systems use a pressure change in the water supply to activate the valves. The portable lateral, buried lateral, and sequencing valve lateral systems can be automated by the use of electric or pneumatic valves which are activated by controllers. These automatic controllers can be programmed for irrigation, crop cooling, and frost control and can be activated by soil moisture measuring and temperature sensing devices.

Gun and boom sprinklers

Gun sprinklers have 5/8 inch or larger nozzles attached to long discharge tubes. Most gun sprinklers are rotated by means of a rockerarm drive and most can be set to irrigate a part circle.

Fig Part-circle gun sprinkler with rocker-arm drive

Boom sprinklers have a rotating 110-250-foot boom supported in the middle by a tower mounted on a trailer. The tower serves as the pivot for the boom that is rotated once every 1 to 5 minutes by jets of water discharged from nozzles. The nozzles are spaced and sized to apply a fairly uniform application of water to a circular area over 300 feet in diameter. Tower movement is periodic and manual.

Gun or boom sprinkler systems can be used in many similar situations and are well adapted to supplemental irrigation and for use on irregularly shaped fields with obstructions. Each has its comparative advantages and disadvantages. Gun sprinklers are considerably less expensive and are simpler to operate; consequently, there are many more gun than boom sprinklers in use. However, guns require high pressure and therefore energy costs are higher. Gun and boom sprinklers usually discharge more than 100 gallons per minute and are operated individually rather than as sprinkler-laterals.

Gun and boom sprinklers can be used on most crops, but they produce relatively high application rates and large water drops that tend to compact and seal the soil surface and create runoff problems. Therefore, these sprinklers are most suitable for coarse-textured soils having high infiltration rates and for relatively mature crops that need only supplemental irrigation.

Gun and boom sprinklers are usually not recommended for use in extremely windy areas because their distribution patterns may become too distorted. Large gun sprinklers are usually trailer or skid mounted and like boom sprinklers are towed from one position to another by a tractor. A disadvantage of boom sprinklers is that they are unstable and can tip over when being towed over rolling or steep topography.

2.5.2 Continuous-move sprinkler systems

Center-pivots

Center pivot systems sprinkle water from a moving lateral pipeline. The center pivot lateral is fixed at one end and rotates to irrigate a large circular area. The fixed end of the lateral, called the pivot point, is connected to the water supply. The lateral consists of a series of spans with steel trusses ranging in length from 90 to 250 feet and carried about 10 feet above the ground by drive units.

Fig Outer end of a center-pivot lateral in operation

A drive unit consists of an A-frame supported on motor-driven wheels. The last downstream drive unit moves continuously or intermittently to set the rotation speed, and all other drive units move intermittently to maintain the lateral pipe in an approximately straight alignment. The most common center-pivot lateral uses 6 or 6-5/8 inch pipe, is nearly a quarter of a mile long, and irrigates the circular portion of a quarter section.

However, laterals as short as 220 feet and as long as half mile 800 m are available. Laterals longer than a quarter mile generally have one or more initial pipe spans of 8 or even 10 inch pipe. A guidance system composed of devices installed at each drive unit keep the lateral in a line between the pivot and end-drive unit, and the end-drive unit is set to control the speed of rotation. Mainlines supplying center pivots are buried and may be comprised of networks that supply multiple center pivot systems or single pipelines connecting the water source to single center-pivot system.

The moving lateral pipeline is fitted with sprinklers to spread the water evenly over the circular field. The area to be irrigated by a nozzle along the lateral becomes progressively larger toward the moving end, and the lateral speed becomes progressively faster.

Therefore, to provide uniform application, the sprinklers must be designed to have progressively greater discharges, closer spacing, or both, toward the moving end. Typically, the application rate near the moving end is about 2.5 inches per hour. This exceeds the intake rate of many soils except during the first few minutes at the beginning of each wetting event. To minimize runoff, the laterals are usually timed to rotate once every 14-84 hours depending on the soil's infiltration characteristics, the system's capacity, and the maximum desired soil moisture deficit.

Five types of power units commonly used to drive the wheels on center pivots are electric motors, hydraulic oil motors, water pistons, water spinners and turbines, and air pistons. The first pivots were powered by water pistons; however, electric motors are by far most common today because of their speed, reliability, and ability to run backwards and forwards. Electric and hydraulic oil motors allow the system to be operated dry.

Self-propelled, center-pivot sprinkler systems are suitable for almost all field crops, but require fields free from any obstructions above ground, such as telephone lines, electric power poles, buildings, and trees. They are best adapted for use on soils having high intake rates and on uniform topography. When used on soils with low intake rate and irregular topography, any runoff can cause erosion and puddles that may interfere with the uniform movement of the lateral and traction of wheels. If these systems are used on square fields, some means for irrigating the four corners must be provided, or other uses made of the area not irrigated. In a 160-acre quarter-section field, about 30 acres are not irrigated by the center-pivot system unless the pivot is provided with a special corner irrigating apparatus. With some corner systems only about 8 acres are left unirrigated.

Most pivot systems are permanently installed in a given field. However, in supplemental irrigation areas or for dual cropping, it may be practical to move a standard quarter-mile center-pivot lateral back and forth between two 130 acre fields. Many modern center pivot systems are equipped with remotely operated controls that use radio transmission and often internet delivery. Some systems provide audio and visual alarms upon system or water supply malfunction, including calling operators via radio or mobile phone.

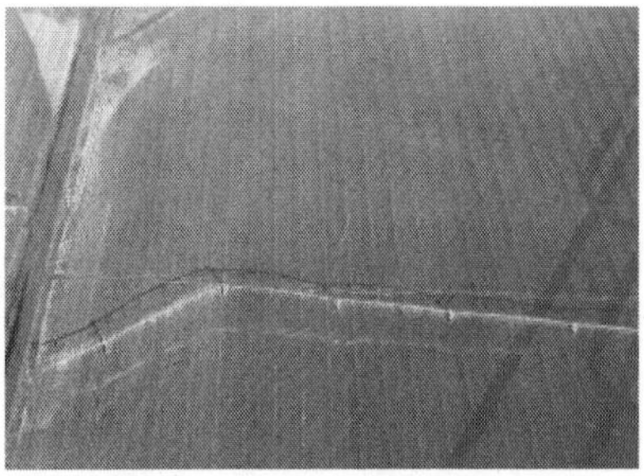

Fig Corner system in operation on a center pivot

Traveling sprinkler

The traveling sprinkler, or traveler, is a high-capacity sprinkler fed with water through a flexible hose, mounted on a self-powered chassis, and travels along a straight line while watering. The most common type of traveler used for agriculture in the United States has a 150 to 500 gallons per minute gun sprinkler that is mounted on a moving vehicle and wets a diameter of more than 400 feet. The unit is often equipped with a water piston or turbine-powered winch that reels in the cable. The cable guides the unit along a path as it tows a flexible high-pressure hose that is connected to a pressurized water supply system. The typical hose is often 4 inches in diameter and approximately 660 feet long, allowing the unit to travel up to 1,320 feet unattended.

Fig Hose-fed traveling gun type sprinkler

In some systems, after use, the hose can be drained, flattened, and wound onto a reel. In other cases the plastic hose does not flatten when empty, and the reel assembly is parked at one end of the field and a tractor or truck pulls the gun-sprinkler carriage to the other end of the field, in a straight line. Then, the water is turned on and the reel slowly rotates, pulling the gun sprinkler in. When the carriage the gun is mounted on arrives at the reel assembly, a mechanism shuts off the system. The unit is moved to the next row and the process repeats until the entire field area is irrigated. The reel assembly rotates using a hydraulic motor which is driven by the water entering the unit, reducing the pressure substantially. The long hose also causes a significant pressure loss. Thus, sufficient supply-line pressure is needed to operate such a system, and it is most convenient to have the water supply at one end of the field, not in the middle of the field. The operator can use controls to adjust the rotation speed of the reel assembly and apply the desired depth of irrigation water.

Some traveling sprinklers have a self-contained pumping plant mounted on the vehicle that pumps water directly from an open ditch while moving. The supply ditches replace the hose. Travelers can also be equipped with boom sprinklers instead of guns. Boom sprinklers can be fixed or have rotating arms 60 to 120 feet long from which water is discharged through nozzles as described.

Fig Fixed-arm boom assembly with low-pressure spray nozzles for traveler system

As the traveler moves along its path, the sprinkler wets a strip of land about 400 feet wide rather than the circular area wetted by a stationary sprinkler. After the unit reaches the end of a travel path, it is moved and set up to water an adjacent strip of land. The overlap of adjacent strips depends on the distance between travel paths and on the diameter of the area wetted by the sprinkler. A part-circle sprinkler is frequently used so that the dry part of the pattern is over the tow path. That ensures the unit travels on dry ground.

Fig Hand-move sprinklers operation

Figure above shows a plan view of a typical traveling sprinkler layout for an 80 acre field. The entire field is irrigated from 8 tow paths, each 1,320 feet long and spaced 330 feet apart. Traveling sprinklers require the highest pressures of any system.

In addition to the 65 to 100 pounds per square inch required at the sprinkler nozzles, hose friction losses add another 20 to 40 pounds per square inch to the required system pressure. Therefore, travelers are best suited for supplemental irrigation where seasonal irrigation requirements are small, thus mitigating the high power costs associated with high operating pressures.

Traveling sprinklers can be used in tall field crops such as corn and sugarcane and have even been used in orchards. They have many of the same advantages and disadvantages described under gun and boom sprinklers; however, because they are moving, traveling sprinklers have a higher uniformity and lower application rate than guns and stationary booms. The application uniformity of travelers is only fair in the central portion of the field, and 100- to 200-foot-wide strips along the ends and sides of the field are usually poorly irrigated.

Linear-move lateral

Self-propelled linear-move laterals combine the structure and guidance system of center-pivot lateral with a traveling water feed system similar to that of a traveling sprinkler. Linear-move laterals require rectangular fields free from obstructions. Measured water distribution from these systems has shown the highest application uniformity coefficients for single irrigations under both calm and windy conditions.

Some linear-move systems pump water from open ditches, but some are supplied by a flexible hose so that some undulation in ground elevation can exist. Other systems utilize automated coupling mechanisms in which the water is supplied through evenly-spaced hydrant valves along a buried mainline pipe. The automated coupling systems use swing arms with swivel joints to connect to the hydrants as the system moves down the field, and they are suitable for application in undulating topographies, but are much more expensive than linear-move systems that take water from a ditch or a flexible hose. In any case, the guidance system must be robust to avoid shutdowns. Common guidance systems utilize an elevated cable or furrow along one side of the field. Other systems use global positioning satellite technology to guide lateral movement.

Linear-move systems that pump from ditches must have a self-contained pump and motor, usually powered by an on-board diesel engine or motor fed via an auxiliary electrical cable. Hose-pull systems generally obtain necessary water pressure from the water source. An auxiliary electrical cable is often used to provide power for drive towers.

A major disadvantage of linear-move systems, as compared to center-pivot systems, is the problem of bringing the lateral back to the starting position. Since the center-pivot lateral operates in a circle, each irrigation cycle automatically ends at the beginning of the next, but because the linear-move lateral moves from one end of the field to the other, the ending position is the maximum distance away from the starting position; down time results if the machine is returned dry. Sometimes, linear-move systems are operated in both directions, applying half of the maximum soil water deficit during each pass. However, the linear-move system can irrigate all of a rectangular field, whereas center-pivot systems can irrigate only an essentially circular area.

Other sprinkler systems Concerns about the availability and cost of energy has been increasing interest in the use of perforated pipe, hose-fed sprinklers operating on a grid pattern, and micro sprinklers. These systems afford a means of irrigating with low pressure. Often, gravity pressure is sufficient to operate these systems without pumps. Furthermore, inexpensive low-pressure pipe such as unreinforced concrete and thin-wall plastic can be used to distribute the water. These systems do have the disadvantage of a high labor requirement if moved periodically rather than being fixed in place.

Perforated pipe

This type of sprinkle irrigation is not commonly used for agricultural irrigation, though they can be inexpensive compared to other types of sprinkler systems. Perforated pipe systems spray water from one-sixteenth-inch-diameter or smaller holes drilled at uniform distances along the top and sides of a lateral pipe or hose. The holes are sized and spaced so as to apply water uniformly between adjacent lines of perforated pipeline.

Fig Side-roll sprinkler lateral

Such systems can operate effectively at pressures between 5 and 30 pounds per square inch, but can be used only on coarse-textured soils such as loamy sands with a high capacity for infiltration because the application rate is quite high. The application rate tends to change rapidly along the length of the lateral, so the water application uniformity is relatively low, compared to most sprinkler systems.

Hose-fed sprinkler grid

These systems use hoses to supply individual small sprinklers that operate at pressures as low as 5 to 10 pounds per square inch. They can also produce if the sprinklers are moved in a systematic grid pattern with sufficient overlap. However, these systems are not in common use except in home gardens and turf irrigation, although they do hold promise for rather broad use on small farms in developing countries where capital and power resources are limited and labor is relatively abundant.

Orchard sprinklers

Orchard sprinklers are small spinners or impact sprinklers designed to cover the space between adjacent trees with little or no overlap between the areas wetted by neighboring sprinklers.

Orchard sprinklers operate at pressures between 10 and 30 pounds per square inch (70 and 200 kPa), and typically the diameter of coverage is between 5 and 30 feet (2 and 10 m). Discharge rates range from 5 to 100 gallons per hour (20 to 375 lph). Orchard sprinklers are located under the tree canopies to provide approximately uniform volumes of water for each tree, but without wetting the tree foliage.

Water should be applied fairly evenly to areas wetted, although some soil around each tree may receive little or no irrigation as shown. Individual sprinklers can be supplied by aluminum laterals or hoses and periodically moved to cover several positions or an orchard sprinkler can be provided for each position on lateral lines that are similar to those used for drip irrigation. Orchard sprinklers are typically fixed to reduce labor costs and to provide some degree of frost protection.

2.6 Operational requirements for sprinkler irrigation system

Sprinkler irrigation system should be operated in keeping with good irrigation practices. It should be ensured that the prime mover and the pump are in alignment. For these the drive shaft as well as the pump shaft should lie at nearly the same height to prevent too great an angle on the universal shaft.

While laying the main and lateral pipes, always begin laying at the pump. This necessarily gives the correct connection of all quick coupling pipes. While joining couplings, it is ensured that both the couplings and the rubber seal rings are clean. In starting the sprinkler system, the motor or engine is started with the valves closed. The pump must attain the pressure stated on type-plate or otherwise there is a fault in the suction line. After the pump reaches the regulation pressure, the delivery valve is opened slowly. Similarly, the delivery valve is closed after stopping the power unit. The pipes and sprinkler-lines are shifted as required after stopping. Dismantling of the installation takes place in the reverse order to the assembly described above.

The principle of operation of the hand-move system is based on the movement of the laterals from one position to the next after a predetermined irrigation event. Each irrigation event has a set time that depends on the amount of water required by the crop at that stage of growth. For the semi portable irrigation system, the lateral is coupled directly to the valve control, elbow or to a header, which intern coupled directly to the valve control elbow. After irrigating in one position, the lateral is uncoupled and moved to the next position. Pipes should not be dragged along the ground as this result in damage and besides that soil would enter the pipes. Care should be also taken that when the pipes are moved is an area with electricity lines, there is a no contact with the lines. For drag-hose sprinkler systems, the hose is lateral. The hose is connected with garden tap or turf hydrant at one end and the riser on the other end. Hose clips and adapters are used to secure the hose. After irrigating one position, all sprinklers, risers and tripod are moved to the next position. Of particular importance to this system is that users should clearly mark on the hose all position of the sprinkler, so as to ensure that they place the sprinkler at the same position each time that position is irrigated and also to ensure that the overlap envisaged in the design is maintained.

Chapter 3

Drip Irrigation System

3.1 Introduction

Drip irrigation is a method of irrigation where water is dripped directly to the root zone of plants which gets slowly delivered to crop root zone by eliminating runoff, evaporation and drift. A well designed and managed drip irrigation system gives the producers the best uniformity and application efficiency available. It is also known as micro irrigation or localized irrigation.

Drip irrigation involves supplying water to the soil very close to the plants at very low flow rates from plastic pipe fitted with outlets. The basic concept underlying the drip irrigation method is to maintain a wet bulb of soil in which plant roots suck water. Only the part of the soil immediately surrounding the plant is wetted. The volume and shape of the wet bulb irrigated by each drip emitter are a function of the characteristics of the soil and the discharge rate of the drip emitter.

Drip Irrigation or trickle irrigation is an irrigation method which saves water and fertilizer by allowing water to drip slowly to the roots of plants, either onto the soil surface or directly onto the root zone, through a network of valves, pipes, tubing, and drippers.

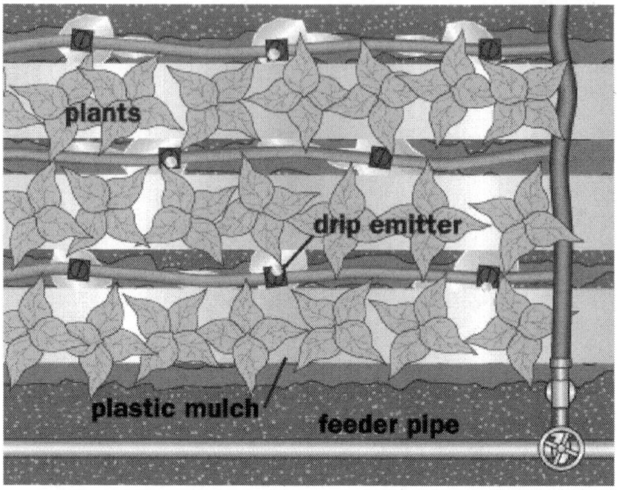

In this, water and nutrients are delivered across the field in pipes called as dripper lines featuring smaller units known as drippers. Each dripper emits drops containing water and fertilizers, resulting in the uniform application of water and nutrients direct to each plant's root zone, across an entire field. It provides water near the base of the plant, leaving the upper foliage dry and less susceptible to fungi.

3.2 Benefits of drip irrigation

Drip irrigation allows water to trickle slowly near the base of the plant by system of tubing, pipes, valves and emitters. Many systems are relatively easy to set up, and once in place, drip irrigation creates healthier growing conditions and higher yields while saving time and money for gardeners.

There are many benefits of using drip irrigation system, which includes:

- ❖ Better plant health: In this, water penetrates slowly and deeply into the soil, placing moisture exactly where it's needed at the roots. With this, plants grow quickly when are watered uniformly without stress created by lack of water.
- ❖ Less waste: The drip irrigation system uses one-quarter to half the amount of water used by traditional overhead sprinklers. In this, 70% of water ends up in the soil, and the rest is lost to evaporation and wind. In case of good drip irrigation system, there will be 90% of water in soil with very little loss to evaporation.
- ❖ Environmental benefits: As drip irrigations are effective, there is less chance of erosion, with very little runoff ending up in area streams and rivers.
- ❖ Reduced risk of disease: Similar to overhead sprinklers, drip irrigation system also keeps the foliage dry and reduces incidence of powdery mildew and other diseases that occur in damp conditions.
- ❖ Money savings: A good drip irrigation system is more effective where monthly water bill appears to be smaller.

- ❖ Saves time: With drip irrigation system, there is no need to continually drag hoses and sprinklers from one spot to another. Also many irrigation kits make it easy to install a timer, which provides level of automation that makes irrigation even easier.
- ❖ Simplicity: High quality drip irrigation are easy to assemble and even easier to use. Such systems are attached directly to the hose spigot.
- ❖ Weed control: Drip irrigation systems place water to the area directly around the plant. With this, weed seeds are water starved and germination is limited.
- ❖ Adaptability: When properly configured, drip irrigation system can work on any soil type, and on unleveled landscapes.
- ❖ Appearance: The appearance of a drip irrigation system is simple where tubes can be hided with layer of organic mulch.

3.3 Limitations of drip irrigation

1. Initial set up cost is high.
2. Direct sunlight reduces the life span of the tubes used for supplying water.
3. Problems in moisture distribution.
4. Improper filtration may cause clogging.
6. Salinity problem.
7. Requires high skill for the installation and maintenance.
8. Flushing of tubes is required periodically.

3.4 Components of drip irrigation system

Drip irrigation system, also known as trickle irrigation, involves dripping water on the soil at very low rates, and thus, provides a very favourable high moisture level in the soil in which plants can flourish. Drip irrigation is most suitable for row crops (vegetables, soft fruit), tree and vine crops where one or more emitters can be provided for each plant. Because of the high capital costs of installation, only high value crops are grown in drip system. A drip irrigation system is comprised of several components.

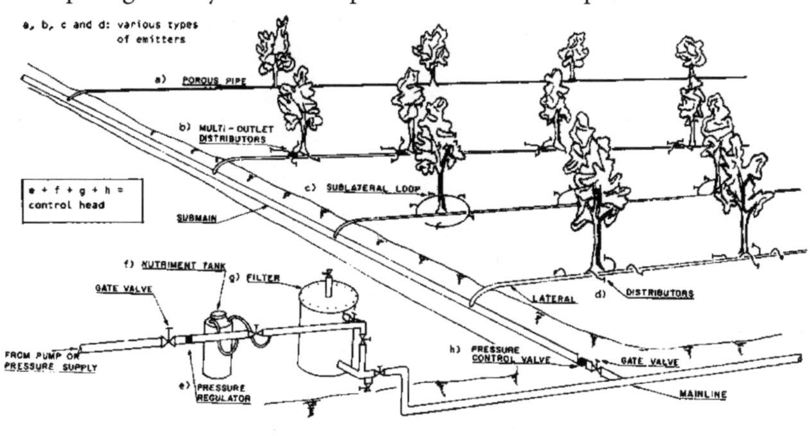

The basic components of any drip irrigation system are broadly divided into the following components:

- ❖ Pump unit
- ❖ Control head
- ❖ Mainlines, Submains and Laterals
- ❖ Emitters or drippers

I. Pump Unit:

It consists of prime mover (motor/engine)/water pump, which takes water from the source and provides the right pressure for delivery into the pipe system consisting of G.I. pipes.

A pump of suitable capacity is used to supply water through the components of drip irrigation system at specific level of pressure. If the source of water supply is bore well, open well or canal, there is the possibility of organic and inorganic foreign bodies in the water. In this case, suction filter is to be used to get relatively clean water.

The electric motors or diesel engines are the common prime mover of the pump. Also, the solar pump is being used to popularize it for drip irrigation purposes.

II. Control Head:

Fig Filter

It consists of bye pass and air release valves to control the discharge and pressure in the entire system. It consists of filters to clear the water such as screen filters and graded sand filters, which remove fine material suspended in the water. Screen filters or disk filters may be used with groundwater. A 200-mesh screen or equivalent is considered adequate for drip irrigation.

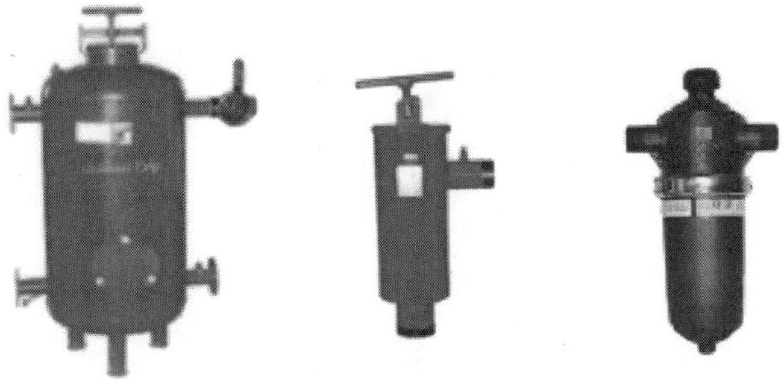

Sand filter Screen filter Disc filter

When the water contains sand, a sand separator should be used. Rapid clogging may occur when no filter or the incorrect type of filter is used. A filter needs to be cleaned when the difference in pressure across the filter is greater than 5-8 psi. A drip-irrigation system should never be operated without a filter even if the filter requires frequent cleaning. Failure to use a filter will result in clogging of the emitters, often resulting in poor uniformity and sometimes in crop loss.

The filter should be cleaned as often as needed. Efforts should be made to understand the cause of the rapid clogging, and remediation for the problem should be developed. The presence of the filter after the point of fertilizer injection means totally soluble fertilizers must be used. Otherwise, fertilizer particles may contribute to filter clogging. Some control head units contain a fertilizer or nutrient tank.

Fig Fertilizer tank

These slowly add a measured dose of fertilizer into the water during irrigation through injectors. This is one of the major advantages of drip irrigation over other methods. The most common injectors used with small drip-irrigation systems are the Venturi injector and the Dosatron. Because Venturi injectors involve no moving parts and are less expensive, they are commonly used on small farms and located as close as possible to the irrigation zone, but before the filter.

III. Mainlines, Submains and Laterals:

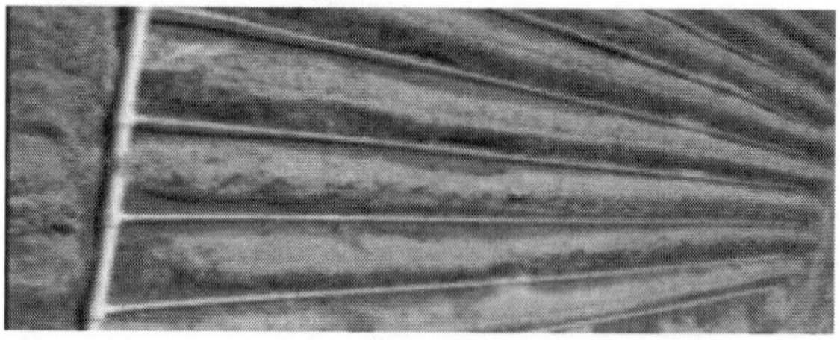

Fig Main and Sub lines

These water lines supply water from the control head into the fields. They are usually made from PVC or polyethylene hose and should be buried below ground because they easily degrade when exposed to direct solar radiation. The mainline is the pipe that goes from the water source to the control valves which are made of galvanized steel, copper, PVC, heavy wall polyethylene or PEX.

The sub-main feed to the laterals on one or both sides. It is made of either of medium density polyethylene (PE) or PVC. There should be a balance between the diameter of the main and sub-mains.

The lateral is the pipe located between a control valve and the drip tube which are normally 13-32 mm diameter and are provided with end plug and pressure gage. The lateral pipe may be PVC, PEX, or polyethylene.

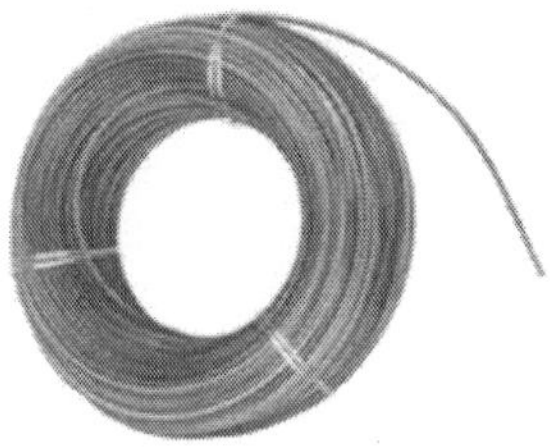

The lateral is located after the pressure regulator so it is not necessary to use a pipe with a high pressure rating. Laterals are often used when multiple drip tubes are needed, such as when the irrigated area is too large for a single drip tube.

IV. Emitters:

Fig Dripper/Emitter

Emitters, the core of micro irrigation system or made of plastic material. The design of production of high quality drippers is comprised of delicate and complicated process. Water passes through the emitters and need to be delivered at constant and low with the desired uniformity. The emitters are designed to dissipate pressure and yield low discharge which does not vary significantly because of minor differences in pressure head. Emitters are classified on the basis of various characteristics:

- Flow regime
- Pressure dissipation
- Operating pressure
- Discharge
- Lateral connection
- Water distribution
- Flow cross section
- Cleaning characteristics
- Pressure compensation
- Material used for production

Drippers

These are also called drippers. These are the devices used to control the discharge of water from the lateral to the plants. These are usually spaced more than 1 metre apart with one or more emitters used for a single plant such as a tree. For row crops, more closely-spaced emitters may be used to wet a strip of soil. Many different emitter designs have been produced in recent years. The basis of design is to produce an emitter which will provide a specified constant discharge, which does not vary much with pressure changes, and does not block easily.

Mainly, there are following types of dripper available;

- Point Sources Dripper: In this system drippers are mounted are inserted along the laterals at length intervals that create a discrete wetted soil volume by each emitter without overlapping.
- Line Sources: Drippers are densely poisoned along the lateral, insuring overlapping of wetted soil volumes by execute and drippers. This layout is typical end tape designed and is the fevered choice of densely grown annual crops.
- Compensating Emitter: Several methods of pressure dissipation are employed in emitter or dripper manufacturing in order to overcome the opposing constraints imposed by energy dissipation and clogging. Pressure compensating emitters are designed to discharge water at a constant rate over a wide range of operating pressures.
- Continuous Flushing Emitter: These are designed to continuously permit passage of large solid particles while operating at a drip flow thus reducing filter finesse requirements.

- ❖ Flushing Emitter: Water is discharged from closely spaced perforations, emitters, or porous wall along the tubing.
- ❖ Long Path Emitters: The water flows through a narrow, long micro tube. The micro tube may be long (spaghetti) or built in- spiral capsulation. Water flow is laminar in the spaghetti and the semi turbulent in the built in spiral. The flow laminar flow dripper is sensitive to change in pressure. The employ a long capillary-sized tube or channel to dissipate pressure.
- ❖ Multi-Outlet Emitters: Each emitter has to hyphen 12 outlets on to which small diameter micro tubes are connected. The drippers are used mostly in landscaping and irrigation of potted plants. Supplies water to two or more points through small diameter auxiliary tubing.
- ❖ Orifice Emitter: Employs a series of orifices to dissipate pressure. Pressure description occurs at a tinny inlet in the bottom of the dripper, rendering it proven to plugging.
- ❖ Vortex Emitter: In vortex drippers, water enters tangentially in to a circular chamber, creating a spiral whirlpool that generates high head losses along a relatively short path. This allows for a wide hyphen orated orifice that decreases clogging hazards. The desirable features of an emitter are low in cost, easy to manufacture and install, resistant to clogging, uniformity in discharge and reliable performance. Emitters are located at predetermined spacing on the lateral and are connected by various means. On the basis of their design, drippers are classified into three types, namely, point source, line source and disc source.

3.5 Types of drip irrigation systems

Drip irrigation is the practice of applying small amount of water and fertilizer uniformly across a specific area. In this, water and fertilizer are delivered directly to the crop root zone, which eliminates runoff, evaporation and drift. A properly designed and managed drip irrigation system gives producers the best uniformity and application efficiency available, consequently saving them time, energy, and water, all while maximizing yields. There are two main types of drip irrigation systems; surface and sub-surface drip systems.

3.5.1 Surface drip system

Surface drip irrigation is also known as drop-by-drop irrigation, trickle irrigation, micro irrigation or localized irrigation which consists of polyethylene pipe, which is implanted by pressure compensating dripper.

Surface Drip

In this system, the water is slowly delivered to the surface of the soil above the root zone of plants. This type of irrigation is mostly used on high-value crops. In this, emitter and lateral pipe are placed on the soil surface. It is the most common and popular type of drip system suitable for wide-spaced plants as well as for row crop. It is easy to use a surface drip to observe and inspect, change and clean the emitters, observe surface moisture patterns and measure individual emitted discharge rates.

Advantages

- Because it is so widely utilised, local irrigators generally have at least minimal understanding of how to operate and maintain the system
- Surface irrigation systems can be developed at the farm level with minimal capital investment
- The essential structural elements are located at the edges of the fields, which facilitates operation and maintenance activities
- If the topography is not too undulating, these costs are not great
- Energy requirements for surface irrigation systems come from gravity
- Surface irrigation systems are less affected by climatic and water quality characteristics
- The gravity flow system is a highly flexible, relatively easily managed method of irrigation

Disadvantages

- The soil, which must be used to convey the water over the field, has properties that are highly varied both spatially and temporally
- Surface irrigation systems are typically less efficient in applying water than either sprinkler or trickle systems
- The need to use the field surface as a conveyance and distribution facility requires that fields be well graded if possible

❖ Surface systems tend to be labour-intensive

3.5.2. Sub-surface drip system

Subsurface drip system is an innovative solution which is powered by latest generation of polyethylene tapes that can be buried to a depth of about 30 cm. These ducts are fitted with drippers at a space of around 40-50 cm, with flow rate, that remains closed to nominal speed and with good uniformity, when operating in variation of pressure range (0.5- 2.5 bar).

The innovative nature of this equipment is that it opposes the invasion of drippers by particles of soil and roots. In this, the emitters are provided with anti-siphon system and tape empties automatically as soon as water gets cut. Such features differentiate the system from the systems tested more than twenty years ago as porous pipes.

Subsurface Drip

Subsurface drip is a highly efficient irrigation system that uses buried drip tubes or drip tape to meet crop water needs. Since the water is applied below the soil surface, the effects of surface infiltration, such as crusting, saturated condition of ponding water, and water losses via evaporation and surface runoff are eliminated. With an appropriately sized and well-maintained subsurface drip irrigation system, water application is highly efficient and uniform.

In this, wetting occurs around the tube and water moves out in all directions. The water is applied directly to the root zone of the crop as opposed to the soil surface where most weed seeds hibernate. As a result, germination of annual weed is reduced. This lowers the pressure on valuable crops. Furthermore, some crops may benefit from the additional heat provided by dry surface conditions, and produce more biomass. When managed properly the application of fertiliser can be optimised. Fields can still be worked when irrigation systems are installed.

Due to sub-surface drip method, there is less intervention in agriculture or any cultural practices and possibly more operational life. Sub-surface drip system provides little interference with cultivation or any cultural practices and possibly longer operational life.

Design

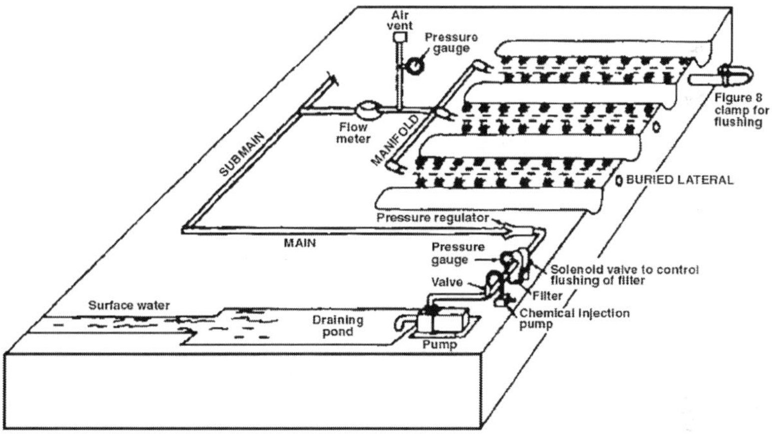

A subsurface drip irrigation system has similar design as a common drip irrigation system. A typical system layout consists of:

- settling pond,
- pumping unit,
- pressure relief valve,
- check valve or back flow prevention valves,
- hydrocyclone separator,
- chemical/fertiliser injection unit,
- filtration unit equipped with back flush valves,
- pressure regulators,
- air vent valves
- PVC pipes to deliver water to the crop.

In this, the piping is 10 to 60 cm below the ground, depending on crop and soil. As a water source, treated grey water or even black water is possible, with the risk of clogging being greater if the influent flow has not properly settled, so treatment of water before the settling pond is necessary.

Subsurface drip irrigation is generally a high-tech, automatically operated technology applied to medium and large scale production. However, several low-cost and simple methods of subsurface drip irrigation like pitcher or bottle irrigation exist that are equally effective for small-scale farming. There are several subsurface techniques used for secondary wastewater treatment such as leach field or evapotranspiration bed that also provide uncontrolled irrigation to fields.

Performance

The performance and life of such system depends on how well it is designed, operated and maintained. It should either be automatically controlled or regularly inspected. Repairing the buried pipes is difficult, cumbersome and time-consuming. To prevent rodents from chewing the pipes, precautionary measures should be undertaken. The mechanical components such as pumps, valves and filters need to be maintained as well as checked and cleaned regularly.

In this, filter system facility is complex and its proper functioning is crucial to the whole system. The maintenance of three main filters of the system should be carried out carefully. Systems must be designed so that mainlines, sub-mains, manifolds and laterals can all be flushed. Mainlines, sub-mains and manifolds are flushed with a valve installed at the very end of each line. Lateral lines can be flushed manually or automatically. It is important to flush the lines at least every two weeks during the growing season.

Subsurface drip irrigation systems are highly efficient irrigation systems that apply accurate amounts of water directly to the root zone, preventing water loss through evaporation and other negative effects of surface irrigation. This system is especially suitable for arid, semi-arid, hot, and windy areas with limited water supply. Since the system is complex and automated, it is more suitable for medium to large-scale production.

Advantages

- High degree of control over water application with the potential for high uniformity of application
- Evaporation is reduced
- The amount of water can be fine-tuned. This avoids water loss caused by run off or evaporation
- Frequent irrigation allows for optimum soil moisture content in the root zone
- Great performance in windy and arid locations
- If pre-treated wastewater is used for irrigation, the risk of direct contact with crops and labourers is reduced

Disadvantages

- Risk of clogging
- When saline water is used, salts accumulate at the wetting front

- ❖ Emitter can be damaged or blocked by root hairs
- ❖ Bacterial slimes and algae growing on the interior walls of the laterals and emitters combined with clay particles in the water can block the emitters
- ❖ Suspended organic matter and clay particles can damage the system
- ❖ A lot of repair work is caused by rodents chewing the tubes
- ❖ Heavy machinery can damage the laterals

3.6 Operational requirements of drip irrigation system

It is important for micro irrigation technician to understand how the drip irrigation system is put together and why certain design elements are specified/required.

Drip line Specifications

The following drip line recommendations serves as guidelines which affects soil type, topography and water quality design. The micro irrigation technician should be familiar with local conditions and recommend steps for correct working of the system. The dripline should be installed with GPS where possible so that its position can be determined as necessary. Depending upon local conditions, dripline can either have pressure compensating or non-compensating drippers. Factors like length of run, topography, zone size and water quality play an important role in selecting the right dripper. Regardless of the type of dripper used, there are several basic guidelines which micro irrigation technician should follow:

- The distance between driplines is determined by the crop. For crops with rows, every 40" the typical spacing between rows of driplines is 80", with dripline located in furrow and one dripline in every other row. With such layout, one dripline irrigates two rows of crop with one irrigated row followed by dry row. Such arrangement has implications for measuring the moisture content of the soil. The micro irrigation technician should take care that measurement should never be taken in dry row. Some growers increases the density of crop plantings by moving to 20" rows and in such case, dripline is placed on 40" centers. In all cases the water application rate is set by design to meet the crop needs given the water availability.
- Driplines are generally buried at a depth of 12" to 18". The crop being grown, the soil texture and the presence of rodents are main considerations for determining the burial depth of the dripline. Sandy soils generally require more shallow burial. In areas with large rodent population deeper burial of dripline is less likely to cross paths with rodent's teeth. In general, rodents are not fond of sandy soil, so shallower depth is not an issue. Deeper placement results in difficulty to germinate the crop with drip system, till there is sufficient residual moisture in the soil. The following should be taken into consideration when designing the system:

- ❖ The distance between drippers and the dripper flow rate should be such selected to achieve the appropriate application rate as per given water availability.

- ❖ Dripline wall thickness should be around 13 to 35 mil which is usually recommended as 15 to 25 mil.

Pump Requirements

The volume output of pumping station shows the amount of area that can be irrigated. A simple formula has been derived converting the maximum required evapotranspiration rate in inches of water per day per acre into gallons per minute per acre which is:

ET (inches/day/acre) *18.86 (conversion factor) = GPM/acre

With this, ET of 0.25 inches per 24 hours per acre would need 4.72 GPM/acre. This calculation is for pump running 24 hours. For safety factor, systems are sized for 20 hours of operation, so for this, micro irrigation technician should use the following formula:

24 (hours in a day)/ (number of hours desired for irrigation) X (GPM/acre)

= 24/20 X 4.72 = 5.66 GPM/acre

On flat land the pressure output required of pump stations is shown by flushing needs of the filters and pipes. On hilly terrain the pressure required to lift water to the highest point must also be considered. Most automatic filters require a minimum of 30 psi to self-clean properly. This is generally the pump's minimum operating pressure to operate a drip system.

Filteration

The filter system protects the drip system from sand and other small particles which can plug the dripline's drippers. A well designed filter system maximizes the performance and longevity of the irrigation system, so for this, there is recommendation for two types of filters:

1. Sand media filters
2. Netafim disc filters

Screen filters are good only for very clean water sources while sand media and disc filters which utilize depth filtration are most effective at removing suspended particles from the water. The filter system should be setup to automatically clean when the pressure differential across the media is too large. A pressure differential switch in combination with a flushing controller is a common approach for automation of filter cleaning.

Pressure Regulating Valves

Pressure control valves are recommended for non-pressure compensated dripline to achieve correct working pressure in drip system. Pressure regulating valves must be adjustable to accommodate higher pressures required during flushing.

Air Vents

Air vacuum vents prevent soil suction into the drippers at system shut-down. For every 50 laterals there should be one anti-vacuum vent at the highest elevation and one mounted on the flushing manifold's highest elevation. A double-purpose automatic air vent must be installed at the pump and is usually required in the mainline.

Fertilizer Injection System

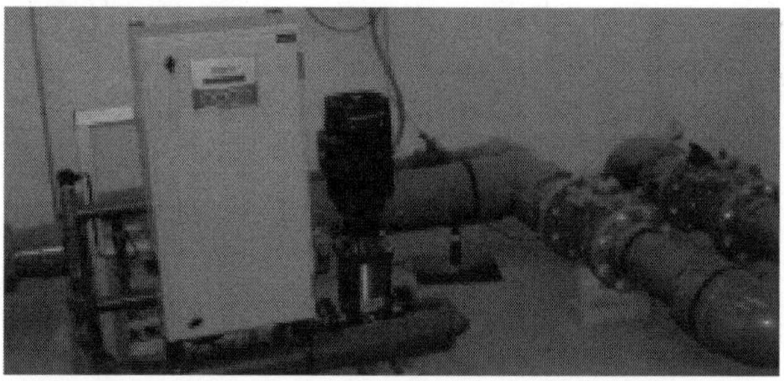

The injection system is designed to supply fertilizer to all irrigation blocks using either an automated system or simple injection pump. For this, micro irrigation technician has to consult with required dealer in order to determine, which fertilizers may be safely applied through the drip system.

Water Meters

It is essential for micro irrigation technician to monitor flow in order to monitor the operation of the irrigation system and crop's water use. The drip system is designed to produce a specific flow rate at given pressure. Changes in flow rate may indicate leaks in the system, improperly set pressure regulating valves or even changes in the well and pumping plant, so micro irrigation technician has to make use of water meter and pressure gauge so as to determine the drip system problems.

Pressure Gauges

The micro irrigation technician should use pressure gauges in order to ensure that the drip system, filter system and pump operates at correct pressure. Pressure gauges are also critical to assess potential problems with the system.

Flush Manifolds

Most permanent drip systems uses flush manifolds to flush entire zones at a single time. A manifold at the end of the field also improves system uniformity, so micro irrigation technician should use flush manifolds to reduce the labor required to properly maintain the system.

3.7 Maintenance of drip irrigation system

Drip irrigation is a mechanical system which performs at a specific pressure of water, if the system has to perform well for longer time, then pressure should be properly maintained. The maintenance of drip system centers on identification of factors which can lead to reduction of performance of the system and procedures to mitigate these negative impacts. Factors that can slow or stop flow through the drip system include:

- suspended material,
- chemical precipitation,
- biological growth,
- root intrusion,
- soil ingestion
- crimping of dripline

To ensure maximum system life reduce or eliminate the impact of the negative factors which requires water treatment and systematic program for regular maintenance. Even if good quality materials are used and installed in a scientific manner, there is a possibility of failure, so field maintenance should be proper and regular.

3.7.1 Importance of Maintenance

There are basically two reasons why maintenance of drip irrigation system is important:

- Water is never found in its purest form in nature as it contains some physical, chemical and biological impurities which may block the pipeline, laterals and drippers in the system.
- The function of dripper / emitter is to allow a gradual transition of water flow from high pressure to atmospheric pressure when it comes out through emitter, so as to get discharge in the form of a droplet. For this, the flow of water has to pass through labyrinth, turbulent and minute flow path. There is always a chance of blockage of this flow path due to dirt particles or due to chemical precipitation.

3.7.2 Schedule Maintenance

In order for drip irrigation system to work smoothly and efficiently for years together, it is very essential for micro irrigation technician to maintain the system with great care. For this, a micro irrigation technician/owner should prepare a system maintenance schedule which should be divided into daily, fortnightly, monthly and half-yearly maintenance activities as shown below.

A) Daily Maintenance

In order to get maximum efficiency and optimum results it is necessary for micro irrigation technician to prevent clogging of emitters, sprinklers and laterals. Properly maintained filters will ensure maximum efficiency of irrigation systems, by avoiding clogging, as filtration unit is the heart of irrigation system.

Following activities should be carried out by micro irrigation technician daily:

1) Start the pump and allow the pressure to become stable. Open the drain valves of hydrocyclone and screen filters to remove the debris.

2) Backwash the sand filter.

Backwashing is the process in which flow direction is reversed so that water flows upwards through the sand bed. In this, sand gets lifted up and expands which allows it to release the dirt arrested in it. The dirt is then driven out of the filter by backwash valve. If backwashing is not done regularly, then impurities will get accumulate in the sand bed which reduces the efficiency of the filter and the system does not get water at desired pressure.

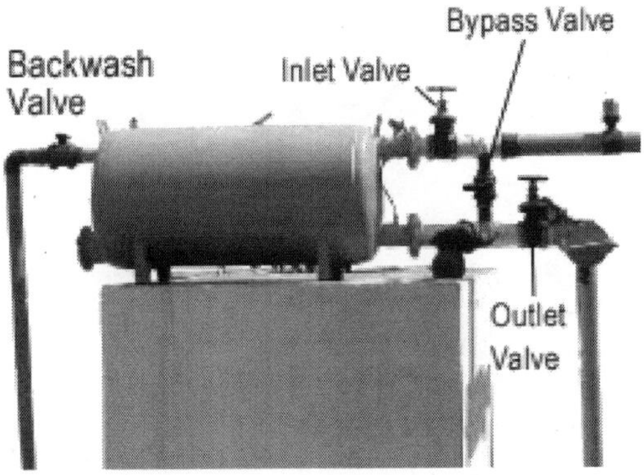

Backwashing of sand filter should be strictly done in the following sequence.

1. Open the Backwash valve.
2. Close the Outlet valve.
3. Open the Bypass valve.
4. Close the Inlet valve.

Backwash operation is complete when clear water starts flowing out through the backwash valve. To resume the filtration process again,

1) Open the Inlet valve
2) Close the Bypass valve
3) Open the Outlet valve
4) Close the Backwash valve.

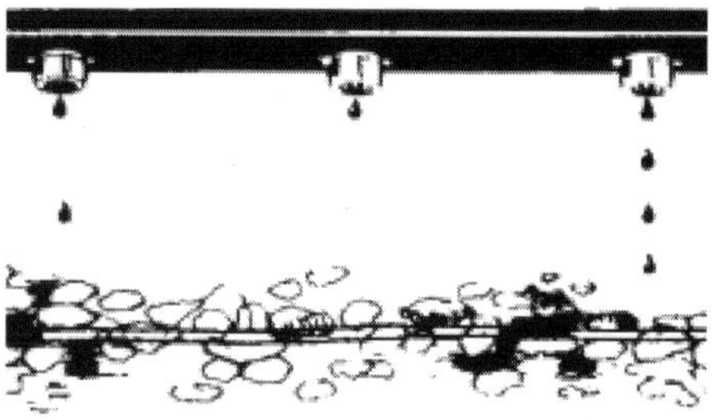

Clean the filters after every 5-6 hours or at recommended timings based on the water quality analysis report.

2) After cleaning the filters, micro irrigation technician has to operate the by-pass valve of the header assembly to get the desired pressure in the system. Operating the system at desired pressure will result in uniform discharge through the drippers and reduction in clogging or choking of laterals and drippers.

3) Micro irrigation technician should take a round of the entire field and should check if there is leakage at joints or damage to any component of the system. If any, then micro irrigation technician should rectify the defects by replacing the spares. Remove the folds or kinks on the laterals, if found and make them straight.

4) Micro irrigation technician should check the drippers for uniformity of discharge. For this, he/she should open and clean the drippers that doesn't emit water. Do not pull the emitter while cleaning it as it will enlarge the hole on the lateral thereby causing leakage.

5) After irrigation is over, micro irrigation technician should check whether the wetting patterns of the drippers are uniform or not. He/she should also check the wet bulb depth taking auger holes at random.

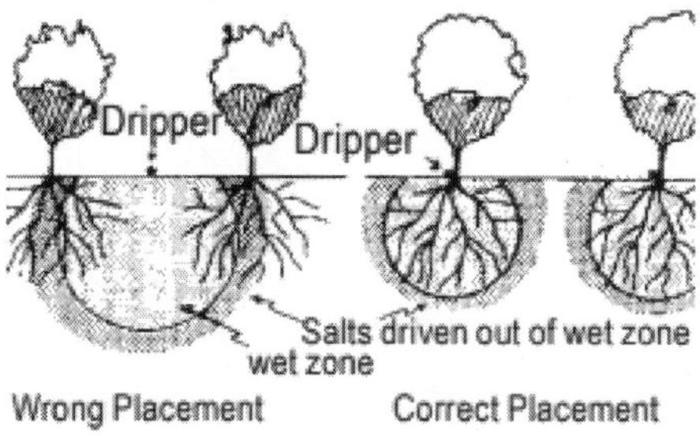

6) Micro irrigation technician should check the position of the drippers, and if the drippers are misplaced; then he/she place them at the correct location. In case of drip irrigated plants, the root development is enormous in wet zone created by drippers and the roots will not get sufficient water which makes the plant growth to hamper if dripper position is changed frequently. In case of saline soil the salts are leached out of the wetted bulb. Micro irrigation technician should change the position of drippers by bringing them back in the root zone which reduces the water uptake rate of plant roots.

7) Remove the end stops and flush the laterals / Inlines thoroughly for about 1-2 minutes. Starting from submain inlet, flush the first 4-5 laterals / Inlines and proceed to the end. This will help in gaining higher velocity in the laterals / Inlines for thorough cleaning.

8) Flush the submain at the end of irrigation to remove the debris. Remember that the impurities accumulate at the submain pipe end and if these are not flushed out from there, these will find access to laterals and clog the drippers. So regular flushing of submains is a must.

B) Fortnightly Maintenance

After completing the steps from 1 to 8 of Daily Maintenance Schedule, micro irrigation technician should perform the following operations at every 15 days interval.

1) Sand Filter

The pressure difference between the inlet and outlet of a filter is an indicator that suggests whether filters need cleaning. If pressure difference is more than 0.5 Kg/cm^2, then filter needs cleaning. The pressure difference between the inlet and outlet of a filter can be checked using a pressure gauge provided with a 3-way control valve.

Due to deposition of salts present in water the top surface of sand bed becomes hard like a stone and backwashing is not effective to clean the sand bed, so micro irrigation technician should clean the sand filter after every 15 days, by following the steps shown:

- Open the lid of sand filter as shown.
- Allow the water to come out through the lid opening.
- Adjust the flow using bypass valve such that sand does not come out of the opening.
- Stir the sand thoroughly by moving the hand through entire sand media from top to bottom. Be careful and do not disturb the position of black filter candles provided at the bottom, else sand may enter the screen filter.
- Break the lumps of sand by squeezing in hand.
- Ensure that half the filter is filled with sand up to the level marked on the filter. Add new sand if it is below the mark.
- Allow water to flow till clean water starts flowing out of the opening.
- Put the lid back in position tightly.

2) Screen Filter

The fine particles and dirt which escape through the sand filter are stopped on the filtering element of screen filter which affects the filtering process. It is essential for micro irrigation technician to clean the filtering element after every 15 days. For doing this, micro irrigation technician should:

- open the lid of screen filter
- take the filtering element out
- remove the rubber seals from both ends of filtering element
- reverse them and clean with water
- fix them tightly on element again
- rinse the element in flowing water gently with hand and clean it

C) Monthly Maintenance

If the salts, algae and other impurities are present in water and enter in drip irrigation system, then laterals and drippers get clogged and may stop emitting water. In such case, micro irrigation technician should necessarily apply acid and chlorine treatments once in a month or as recommended in water quality analysis report.

- Perform acid treatment to remove precipitated salt from drippers and pipeline network.
- Perform chlorine treatment to remove bacterial slime, algae or other biological contamination.
- Inspect all the component above ground for physical abuse, damage by field machinery, rats, squirrels etc. Do not perform both acid and chlorine treatment simultaneously.

D) Half Yearly Maintenance

- Micro irrigation technician should change the sand of media filter with new one as sand particles get rounded off due to continuous abrasion during the operation.
- Micro irrigation technician should check out the system for wear and tear and should replace the spares whenever necessary.
- If the pump works efficiently, it generates the adequate pressure head and discharge required to operate the system satisfactorily. Micro irrigation technician should provide adequate lubrication to the pump and motor. Micro irrigation technician should also make necessary maintenance of pump as per instructions given by pump manufacturer.

Chapter 4

Designing and Layout of Micro Irrigation System

4.1 Study the basics of crop cultivation like spacing, water requirement, spacing of crop stand, nature of rooting etc.

4.1.1 Crop spacing

Introduction

By observing trees and shrubs growing under natural conditions, it is often found that plants grow widely apart in low rainfall areas, so wide spacing of plantings in arid zones generally should be practiced to avoid competition for soil moisture.

The amount of water available to a tree or shrub in a plantation is proportional to the stand density. On dry sites, it is necessary to plant widely apart and to remove all competing ground vegetation; this increases infiltration of rainwater and decreases water losses through transpiration by plants and evaporation from the soil. When irrigation or mechanical cultivation is practiced, it is necessary to adjust spacing to the width of the machinery used and to ensure that plants are placed in straight rows. Actual spacing varies with species, site, and the purpose of the forest plantation. In fuel wood plantations, for example, one might prefer closer spacings than employed in other kinds of plantations.

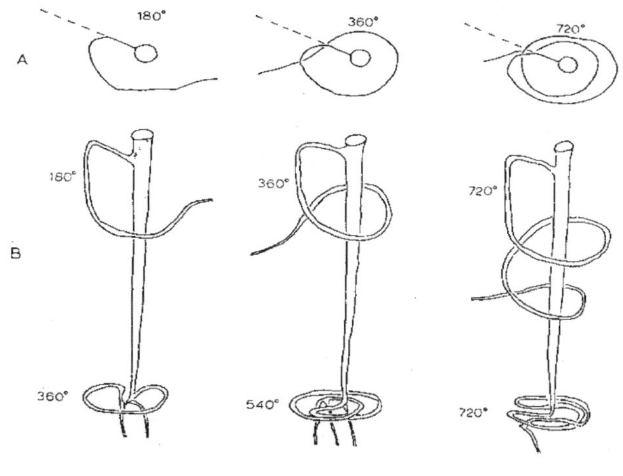

Crop spacing is about the number of crops planted in a unit area. It is about the distance between one plant and another. Micro irrigation technician/agriculturalists have advice about spacing with regard to particular crops based on research findings with view to maximise crop yields. Farmers want to fully exploit their land by getting the best yields and at the same time sustain the productivity of the soil in their gardens. Different annual crops should be well spaced to achieve the best results and there are recommended guidelines.

It is always good to follow the recommended crop spacing guidelines as overcrowding of crops may reduce yields and it may lower the quality of the fruits produced due to competition for light and soil nutrients. If crops are planted too close to one another, it may be hard for the farmer to walk about in the garden weeding, spraying pesticides or doing crop inspection.

When a farmer practices intercropping, it is particularly important to pay attention to spacing. If, for example, maize is to be intercropped with beans, it is good to have wider space between maize plants so that the beans get sufficient light and nutrients to grow well. Similarly, if coffee trees are to be planted in a banana garden, there should be wider spacing for the coffee. For some crops such as millet and some vegetables, the farmer may broadcast the seed at random over the field. This kind of sowing is not done in rows or lines and it is possible for overcrowding to take place in parts of the field which may result in poor yields. The farmer is expected to constantly monitor the crop's growth and to do thinning where possible to ensure correct and uniform plant population. To ease such activities as weeding, harvesting, spraying, and general crop inspection, the farmer is best advised to plant the crops in rows.

Importance of spacing

- Leaving the right amount of space between plants/crops is quite important as each plant/crop needs a certain amount of room for their roots and leaves to maximise growth.
- Leaving enough air to circulate around and in between the plants/croops will really help in preventing diseases from spreading, especially when the weather is wet and humid.
- There are many gardeners who plant closer together than recommended when using raised beds which is based on the idea that raised beds have better quality of soil, so plants can grow better in smaller space.

- ❖ Harvesting is much easier when plants/crops are spaced appropriately as a technician/ gardener can access the crop/plant from all sides without worrying about sabotaging its neighbor, and will able to more accurately assess the yields.
- ❖ Weed management is vital to promote healthy plants, so planting crop/plant too close will make the weeds harder, which ultimately results in more weeding.

4.1.2 Crop Water requirement

Crop water requirements are the depth of water needed to meet the water loss through evapotranspiration of disease free crop, which are grow in large fields under non restricting soil conditions including soil water and fertility and achieving full production potential under the given growing environment.

The crop water need always refers to crop grown under optimal conditions, i.e. a uniform crop, actively growing, completely shading the ground, free of diseases, and favourable soil conditions. The crop thus reaches its full production potential under the given environment.

Factors

The crop water need mainly depends on:

Climate

In sunny and hot climate, crops need more water per day than in cloudy and cool climate. The major climatic factors which influence the crop water needs are:

- ❖ sunshine
- ❖ temperature
- ❖ humidity
- ❖ windspeed

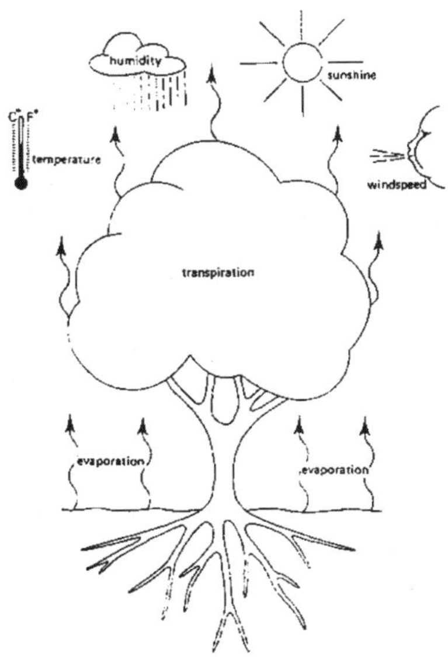

The highest crop water needs are found in areas which are hot, dry, windy and sunny, while the lowest water need are found when it is cool, humid and cloudy with little or no wind. The influence of climate on crop water needs is given by reference crop evapotranspiration which is expressed in millimetres per unit of time, e.g. mm/day, mm/month, or mm/season.

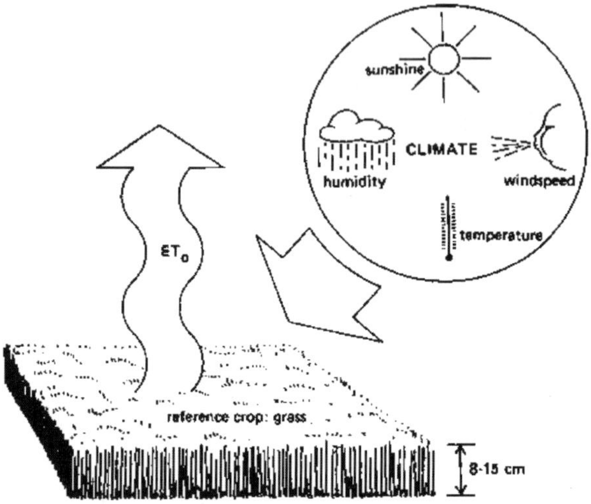

There are several methods to determine the reference crop evapotranspiration which are:

- ❖ experiment using an evaporation pan
- ❖ theoretical using measured climatic data

Experiment using evaporation pan

In this method, pan is filled with water and the loss of water from the pan is measured. Provided that there is no rainfall, the evaporation rate, which is recorded as millimeters per day, is quite easy to measure. This method of measurement takes into account wind, temperature, radiation and humidity, which are the same factors that affect crop transpiration rate.

However, there are a few factors that prevent this recording from being entirely accurate. For one, the solar radiation results in heat storage in the pan. This can lead to increased reading of the evaporation rates at night, when transpiration usually does not occur. In addition, temperature and humidity levels above the pan surface will vary from what would naturally occur.

In this method, different kinds of pans are used to measure the water requirements of crops, with Kp representing the pan coefficient, according to the kind of pan, solar radiation, wind, humidity and the surroundings.

$$ETo = K\ pan \times E\ pan$$

Penman-Monteith Equation

The reference rate, ET_0, is calculated using the Penman Equation, which takes into account the climatic parameters of temperature, solar radiation, wind speed and humidity. variation of this equation, published by the FAO is:

$$ET_o = \frac{0.408\Delta(R_n - G) + \gamma \frac{900}{T+273} u_2(e_s - e_a)}{\Delta + \gamma(1 + 0.34u_2)}$$

Where

ETo reference evapotranspiration [mm day-1],

Rn net radiation at the crop surface [MJ m-2 day-1],

G soil heat flux density [MJ m-2 day-1],

T air temperature at 2 m height [°C],

u2 wind speed at 2 m height [m s-1],

es saturation vapour pressure [kPa],

a actual vapour pressure [kPa],

es - ea saturation vapour pressure deficit [kPa],

D slope vapour pressure curve [kPa °C-1],

g psychrometric constant [kPa °C-1].

Blaney-Criddle Equation

The reference crop evapotranspiration, ET0 can also be calculated by:

$$ET_o = p\ (0.46\ T\ mean + 8)$$

ET_o = Reference crop evapotranspiration (mm/day)

T_{mean} = mean daily temperature (°C)

p = mean daily percentage of annual daytime hours.

As seen from the equation, the method takes into account only the measurement of temperature, so it is not a very accurate estimation of reference evapotranspiration rate.

Crop type

Crops like maize or sugarcane need more water than crops like millet or sorghum. As the influence of climate is given by the reference crop evapotranspiration ETo, so the reference crop used for this purpose is grass. The relationship between the reference grass crop and the crop actually grown is given by crop factor Kc, so the formula:

$$ETo \times Kc = ET\ crop$$

with ET crop = crop evapotranspiration or crop water need (mm/day)

Kc = crop factor

ETo = reference evapotranspiration (mm/day)

Both ET crop and ETo are expressed in same unit as in mm/day or in mm/month. The crop factor Kc mainly depends on:

- type of crop
- growth stage of the crop
- climate

Kc and type of crop

- Fully developed maize, with its large leaf area will be able to transpire and use more water than the reference grass crop: Kc, as maize is higher than 1.
- Cucumber, also fully developed, will use less water than the reference grass crop: Kc, cucumber is less than 1.

Kc and growth stage of crop

- A certain crop will use more water once it is fully developed, compared to a crop which has just recently been planted.

Kc and climate

- The climate influences the duration of total growing period and various growth stages.
- In cool climate, certain crop will grow slower than in warm climate, so to determine the crop factor Kc, it is necessary, for each crop, to know the total length of growing season and lengths of various growth stages.

The determination of Kc values for various growth stages of crops involves:

- Determination of total growing period of each crop
- Determination of various growth stages of each crop
- Determination of Kc values for each crop for each of growth stages

Growth stage of crop

Fully grown crops need more water than crops that have just been planted. Growing period is the period from sowing or transplanting to the last day of the harvest. It is mainly dependent on:

- type of crop and variety
- climate
- planting date

As the growing period heavily depends on local circumstances (e.g. local crop varieties) it is always best to obtain these data locally. Only if no data are available locally should Table 6 be used.

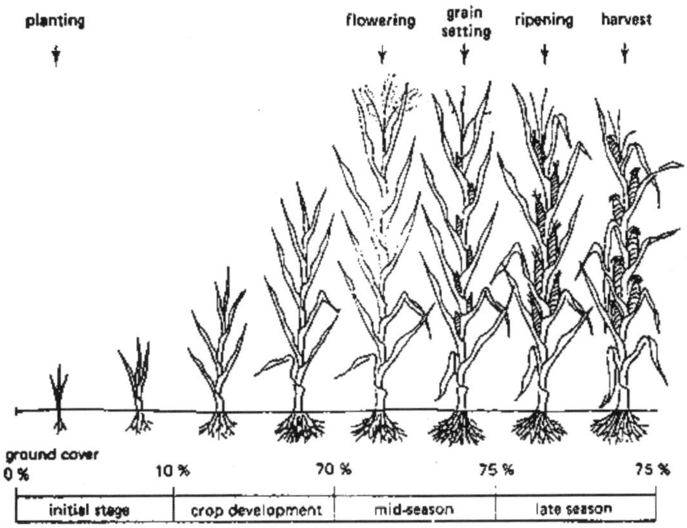

Once the total growing period is known, duration of various growth stages can be determined. The total growing period is divided into 4 growth stages:

- Initial stage: this is the period from sowing or transplanting until the crop covers about 10% of ground.
- Crop development stage: this period starts at the end of initial stage and lasts until full ground cover has been reached. It does not necessarily mean that the crop is at its maximum height.
- Mid - season stage: this period starts at the end of crop development stage and lasts till maturity. It includes flowering and grain-setting.
- Late season stage: this period starts at the end of mid season stage and lasts till last day of harvest. It includes ripening.

Water Requirement of Different Crops

Amount of water required by a crop in its whole production period is called water requirement, so amount of water taken by crops varies considerably.

Crop	Water Requirement (mm)
Rice	900-2500
Wheat	450-650
Sorghum	450-650
Maize	500-800
Sugarcane	1500-2500
Groundnut	500-700

Crop	Water Requirement (mm)
Chillies	500
Sunflower	350-500
Castor	500
Bean	300-500
Cabbage	380-500
Pea	350-500

Crop	Water Requirement (mm)
Cotton	700-1300
Soybean	450-700
Tobacco	400-600
Tomato	600-800
Potato	500-700
Onion	350-550

Crop	Water Requirement (mm)
Banana	1200-2200
Citrus	900-1200
Pineapple	700-1000
Gingelly	350-400
Ragi	400-450
Grape	500-1200

4.1.3 Spacing of crop stand

Introduction

Uniform stand establishment is a critical component in achieving maximum yield potential. With unexpected weather and other uncontrollable factors often impacting the crop's stand, it is even more important to be proactive in the areas that can be controlled.

Factor responsible for high strand

Good crop establishment is one of the most important features in better crop production. The better crop establishment is in turn expressed as optimum plant population in fields. Number of plants per unit area in the cropped field is called as plant population. There are certain steps which will help in getting the crops off to strong start and encourage strong and healthy stand. All of these factors are important because growers only want to plant the field one time. This makes it crucial to get it right the first time.

1. Seed Selection

Each season it all starts with selection of seed. There are multiple considerations which growers need to be aware of when selecting the appropriate seed, which covers geography, unique field conditions, potential impact from weeds, disease and pests, maturity timing, etc. These considerations need to be evaluated in cooperation with farm management practices to ensure the best results.

2. Seed Spacing

Uniform seed spacing is instrument which gives good start to crops. Uneven seed spacing can cost money due to potential yield reduction, so to ensure proper and uniform seed spacing, it's important to make sure that planter is properly calibrated before planting and at the appropriate intervals during planting. As the equipment and field conditions varies, so micro irrigation technician/farmer has to check the equipment manual or with dealer for best recommendations for circumstances. Micro irrigation technician/farmer also want to make sure that the entire planter has been inspected prior to planting and that all the parts are in good working order to make sure that planter is at its best during this very important time.

3. Seed Depth

Planting seeds at uniform seed depth with good seed-to-soil contact is crucial, so micro irrigation technician/farmer has to keep in mind that each crop has its own optimum seeding depth. If seeds are planted too shallow, it may adversely impact the crop's stand

and if roots are not deep enough, they may not be able to support the plant as it matures, which can result in corn plants falling over. It can also result in rootless corn syndrome. When roots are not deep enough, it can also be harder for them to access water in the soil. If seeds are planted too deep, it may take longer for them to emerge, making them more susceptible to disease or pests. Proper down force control during planting is important to maintaining consistent seed depth within a row and within the field. Since field conditions vary, consider your field and make proper adjustments to your planter to ensure optimum seed depth during planting.

4. Uniform Emergence

Uniform emergence also helps achieve better yield. When plants emerge at different rates the smaller plants are at a disadvantage in competing for sunlight and other important resources, but as they try to get the resources to survive, the smaller plants can also negatively impact the larger plants. For best results, it's recommended that all plants emerge within 24-48 hours of each other. To help growers achieve consistent seed spacing, seed depth and uniform emergence within a small planting window, many are relying on precision farming equipment to plant their crops.

5. Fertilizer

In-furrow application is a proven way to ensure the fertilizer gets as close to the seed as possible at planting. With growers planting earlier than ever, fertilizer applied in-furrow will help the seed receive as much support as possible while emerging into cooler or unpredictable environments.

4.1.4 Nature of rooting

Root being the hidden half of the plant, of late emerged as the critical most part for advancing the crop improvement research. While the soil's texture impacts the amount of stored soil moisture, the plant's rooting characteristics determine how much of the soil moisture can be accessed by the plant. A deep rooted crop has access to a greater amount of soil moisture than does a shallow-rooted crop, usually allowing it to go longer between irrigations.

The rooting depth may be affected by the soil depth, constraining soil layers (hardpan, plowpan, etc.), or even abrupt changes in soil texture. When in doubt about the soil profile or rooting depth, use a soil auger or better yet a backhoe, to determine the rooting depth and soil profile textural characteristics.

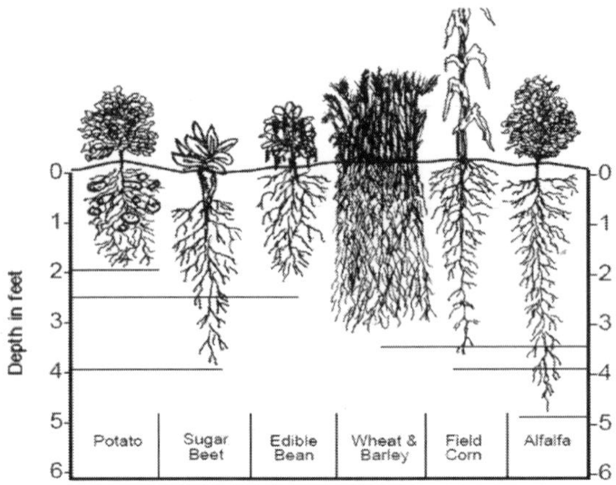

There is an additional dimension to plant-available water capacity of soils: The water in the soil may be available, but roots also need to be able to access it, along with the nutrients contained in the water. The roots could not penetrate into the subsoil and were therefore limited to water and nutrients in the plow layer.

4.2 Familiarize with Sprinklers, Foggers & Misters

4.2.1 Sprinkler

A sprinkler is a device used in irrigation system which irrigates agricultural crops, lawns, landscapes, golf courses and other areas. It is also used for cooling and for the control of airborne dust.

Sprinkler irrigation is a method of applying water in a controlled manner in way similar to rainfall. The water is distributed through a network that may consist of pumps, valves, pipes, and sprinklers. Sprinklers can be used for residential, industrial and agricultural usage. It is useful on uneven land where sufficient water is not available. It is useful for sandy soil. The perpendicular pipes, having rotating nozzles on top, are joined to the main pipeline at regular intervals of time. When water is allowed to flow through the main pipe under pressure with the help of pump it, escapes from the rotating nozzles. It gets sprinkled on the crop.

When choosing a sprinkler or irrigation system, consider selecting one that sends large drops of water close to the ground. These sprinklers are more water-efficient than sprinklers that spray a fine mist and lose a lot of water through evaporation. Drip irrigation is a great alternative for trees, shrubs, perennials and groundcovers. Little or no water is lost to evaporation as the water is applied at ground level, near plant roots.

4.2.2 Types of Sprinkler irrigation

Industrial

- ❖ Higher pressure sprinklers that themselves move in a circle are driven by a ball drive, gear drive, or impact mechanism.
- ❖ These can be designed to rotate in a full or partial circle.

- Rainguns are similar to impact sprinkler, except that they generally operate at very high pressures of 40 to 130 lbf/in^2 and flows of 50 to 1200 US gal/min, usually with nozzle diameters in the range of 0.5 to 1.9 inches.
- In addition to irrigation, guns are used for industrial applications such as dust suppression and logging.
- Many irrigation sprinklers are buried in ground along with their supporting plumbing, although above ground and moving sprinklers are also common.
- Most irrigation sprinklers operate through electric and hydraulic technology and are grouped together in zones that can be collectively turned on and off by actuating a solenoid-controlled valve.

Residential

- Home lawn sprinklers vary widely in their size, cost, and complexity.
- It includes impact sprinklers, oscillating sprinklers, drip sprinklers, underground sprinkler systems, and portable sprinklers.
- Permanently installed systems may often operate on timers or other automated processes.
- They are occasionally installed with retractable heads for aesthetic and practical reasons, reducing damage during lawn mowing.
- These types of systems usually can be programmed to automatically start on a set time and day each week.
- Small portable sprinklers can be temporarily placed on lawns if additional watering is needed or if no permanent system is in place.
- These are often attached to an outdoor water faucet and are placed for a short period of time. Other systems may be professionally installed permanently in the ground and are attached permanently to a home's plumbing system.

Agricultural science

- The first use of sprinklers by farmers was some form of home and golf course type sprinklers.
- Such type of systems are ad hoc systems, while doing the job of the buried pipes and fixed sprinkler heads, interfered with cultivation and were expensive to maintain.

Underground sprinklers

- Underground sprinklers function through means of basic electronic and hydraulic technology.
- This valve and all of the sprinklers that will be activated by this valve are known as a zone.
- Upon activation, the solenoid, which sits on top of the valve is magnetized lifting a small stainless steel plunger in its center.

- By doing this, the activated plunger allows air to escape from the top of a rubber diaphragm located in the center of the valve.
- Water that has been charged and waiting on the bottom of this same diaphragm now has the higher pressure and lifts the diaphragm.
- Such pressurized water is then allowed to escape down stream of the valve through a series of pipes, usually made of PVC or polyethylene pipe.
- At the end of these pipes and flush to ground level are pre measured and spaced out sprinklers.
- Such sprinklers can be fixed spray heads that have a set pattern and generally spray between 1.5–2m, full rotating sprinklers that can spray a broken stream of water from 6–12m, or small drip emitters that release a slow, steady drip of water on more delicate plants such as flowers and shrubs. use of indigenous materials also recommended.

4.2.2 Fogger

Fogging systems are extremely efficient tools for controlling greenhouse climate conditions. As water droplets are released, they quickly convert from liquid to vapor due to their small size and the high temperatures in a greenhouse. During this process, they absorb the hot air at a rate of 16,726 calories per ounce or 590 calories per gram of evaporated water, which helps cool the air temperature and increases humidity levels inside a greenhouse by approximately 6 to 9° F or 4 to 6° C. The overall results and effectiveness of this system depend on outside temperature and humidity levels.

Most fogging systems used for cooling release droplets for a very short time, with approximately 10 seconds between each round of fogging. The duration of a fogging event depends on the air velocity inside the greenhouse. The type of ventilation system installed determines the velocity.

A fogger is used to create a fog, especially for killing insects. They are filled with insecticides and sprayed. A large number of people depend on foggers to drive away pests, owing to its lost cost and high effectiveness. The fogger is ideal for drip irrigation for nurseries, shrubs and trees where a fine mist spray is required at low volumes and low pressures. Foggers can also assist in temperature control and to increase humidity in hot, dry climates. It's rugged nylon construction and easy disassembly allow for carefree operation.

For many reasons fog is a good alternative to mist in specific applications including germination and propagation in greenhouses as fog can provide more uniform wetting across the plants in greenhouse without over-saturation of soil and thus reducing disease and fungal growth. Fog can achieve deeper penetration into the leaves, and many times is more efficient than misting. The real purpose of greenhouse is basically to create a controlled environment for optimal plant growth and development; and in this regard, foggers are also excellent for cooling the environment.

Foggers are recommended for plants that are susceptible to root disease, propagation of seeds and non-rooted cuttings as foggers can better control and build humidity, and keep the root area and leaves the moist without completely saturating the soil.

Foggers are used in overhead applications, generally mounted on or hanging from the overhead supply line of a greenhouse, nursery or hoop-house.

Oversaturation is a common cause of plant diseases, such as fungi and moss, and a danger to greenhouse growers attempting propagation. Foggers can create the ideal conditions for plant propagation by reducing air temperatures and increasing humidity levels inside a greenhouse without wetting plant leaves. Due to the Fogger's smaller droplet size, there is also less soil saturation and more oxygen available to growing plants. The Fogger can also be used for foliar feeding, and to apply insecticides and fungicides. It saves time and provides a uniform application.

4.2.3 Misters

Misters are good way to water plants, to add moisture to air and to cool off in summer. Many times, misters are referred to as foggers that are commonly used on ferns, perennials, garden greens and for starts in garden beds.

Mist is often used for temperature control in outdoor settings such as decks, patios, pools, and animal shelters/enclosures, and in some applications of watering. In this, small droplets are perfect for delicate plant foliage like ferns or ground cover, where traditional sprinklers may be too damaging to plant structure.

Misters, when used for cooling purposes, are recommended to be mounted higher up at around 8-12 feet overhead for achieving good results. It is also recommended to install the mister nozzle at a 90-120 degree angle to the ground. Ideally, the purpose of mist is to evaporate before ever hitting the ground.

Misting plants can be done overhead or raised above the plant canopy on riser and is often recommended to be installed at a 70-degree angle, or upright so that the mist can lightly fall on to the plants before the evaporation.

For densely planted areas or maximum cooling effect, it is required to place the misters close together for full, overlapping, even coverage. They are not recommended to be used in windy areas, since small-sized droplets can be easily dispersed in the air, even under low-to-moderate wind conditions and the wind can negatively affect uniformity and performance.

The advantage of the mister is that a relatively large number can operate at a given time. They are recommended to be spaced every 2-4 feet for optimal coverage. Micro irrigation technician has to confirm the tubing or mainline in their system for volume of water to be used/handled, as that will determine the limitations of how many misters can be installed on one line.

4.2.4 Selection of fogger or mister

There are a couple of factors to consider while selecting fogger or mister which includes:

❖ purpose of device

❖ droplet size

❖ volume of water

❖ PSI range

- Temperature/humid control: Micro irrigation technician has to see whether temperature needs to be controlled by using evaporation (misting) or humidity needs to be controlled by fogging.
- Droplet size: Micro irrigation technician needs to see the droplet size which is measured in microns which best suits as per need. He/she has to analyse to use the application that best determine large or small droplets. Further, micro irrigation technician also needs to know the requirements which can be watering the plants, cooling the air or controlling the humidity, since different droplet sizes correspond to different effects.
- Water requirement: Micro irrigation technician should need to know about the volume of water required. He should estimate the amount of water needs to cool an area, watering the plants or controlling levels of humidity as volume of water which is in gallons/hour which is produced by watering device matters.
- Pressure (PSI): Many foggers and misters operate at high pressure range to achieve the required performance, as high pressures create fog or mist like effects. Whether micro irrigation technician have their existing system or will be installing new system, so he needs to ensure that he can provide the required pressure to operate the devices.

4.3 Familiarize with Principles of Micro Irrigation System

4.3.1 Introduction

The basic principles of micro irrigation explain about the major components of micro irrigation systems with advantages and disadvantages. Drip, trickle and spray irrigation, have been supplanted as micro irrigation which includes methods of frequent water application, in small flow rates on or below the soil surface. In this, volume of water is applied directly to the root zone in quantities that approach the consumptive use of plants. The good management of micro irrigation systems with root zone moisture content gets maintained near field capacity all through the season which provides level of water and air balance that is close to best plant growth. In this, nutrient levels are applied with water so system can be controlled precisely. During dry season in humid areas, or in arid climates, micro irrigation can have significant effect on quality and quantity of yield, pest control and harvest timing.

In micro irrigation systems, water is distributed using an extensive hydraulic pipe network that conveys water from its source to the plant. Outflow from the irrigation system occurs through emitters placed along the water delivery pipes in the form of droplets, tiny streams or miniature sprays. The emitters can be placed either on or below the soil surface. Such system are classified by the type of emitter used in the system which can be drip, bubbler, spray jet, and subsurface.

In this, emitters can vary from sophisticated, constant-flow-rate at variable pressure types of devices to very small, simple orifices. A large number of different types of emitters have been developed in attempts to find a perfect one. The main objective is to assure uniformity of water distribution. It is essential that the discharge from the emitter be uniform and that it not change significantly with small pressure variations in the system. At the same time the emitter should be constructed in such a way that it does not clog very easily. The cost and the size are also important. The emitters presently available can be classified as:

- long path emitters
- short orifice emitters
- vortex emitters
- pressure compensating emitters
- porous pipe or tube emitters

In micro irrigation, emitting devices covers bubblers and jet sprays. In this, emitters are placed along laterals which are the final water delivery lines designed for uniform water distribution. The lateral line is generally constructed of flexible polyvinyl chloride or polyethylene hose. It is often placed above the ground but it can be buried. For row crops, in the line-source type of micro irrigation system, a lateral line combines the function of the line and the emitter. These include laterals constructed from porous pipe, twin bore pipe or pipe provided with evenly spaced, built in emitters.

Each manufacturer supplies data giving allowable length of laterals and recommended pressure for particular line source systems. Recommendations are based on the hydraulic characteristics of the product. The emitters, which may be connected to the lateral using a variety of connectors during installation of the irrigation system, are usually placed at predetermined positions, for example at the base of the irrigated tree.

In this, lateral lines are attached to manifolds or submains where manifolds and submains that distribute the water to specific parts of field, are constructed from flexible smooth-walled, non-collapsible, black PE or flexible PVC pipe which can be left on soil surface and can be also constructed out of rigid PVC pipe that is buried beneath the soil surface for protection against sunlight damage and prevention of algae growth in line. The controls for adjustment of flow rate and pressure are usually located in the submain or manifold line along with valves and timing devices for the separate parts of the field.

Also, water is delivered to the field by the main line which is constructed from white PVC pipe that is buried beneath the soil surface for protection against harmful sunlight. The pipe should be properly rated for the particular application and able to withstand the design pressure in the system.

A main control station, often called the "control head", is usually located close to the water supply. A typical control station includes the pump, backflow prevention system, chemical injection system for fertilizers, chlorine or other chemicals, and a combination of different filters. A main line valve and flowmeter are also included in the control head. Micro irrigation systems can be controlled manually or automatically.

The controller is often located next to other components of the control station. It can control the main valve, chemical injection, backflushing of filters, solenoid valves, and other controls located at remote locations in the irrigation system. Depending on the system, all or some of these components can be automated.

4.3.2 Wetting pattern

The wetting pattern is an important factor to consider when designing and managing a drip irrigation system. The dimensions of the pattern are imperative in selecting the right spacing between emitters and the suitable distance between laterals.

As a result of manner in which water is applied by micro irrigation system, only a portion of the soil surface and root zone of the total field is wetted. Water flowing from the emitter is distributed in the soil by gravity and capillary forces creating the contour lines, often referred to as onion patterns. The exact shape of the wetted volume and moisture distribution will depend on the soil texture, initial soil moisture, and to some degree, on the rate of water application. In line-source type of micro irrigation tube, where the emitters are spaced very closely, individual onion patterns connect/creating a continuous moisture zone along the row.

4.4 Design of Micro Irrigation system

4.4.1 Introduction

Micro-irrigation is a low-pressure, low-flow-rate form of irrigation that can reduce the likelihood of overwatering a landscape. Micro-irrigation delivers water directly to the root zone of plants, where water is needed most, as well as more slowly and over a longer period of time, preventing water from running off on the surface and reducing evaporation associated with sprinkler irrigation. By applying water at a low rate over an extended period of time, micro-irrigation allows the water to better infiltrate into the soil. Water applied by higher flow-rate methods, such as spray sprinklers, can form puddles, evaporate, or land in soil beyond plant roots, leading to less efficient watering. It involves small sprinklers installed at the base of a tree, for example in an orchard or vineyard. Water is supplied to the sprinkler either via pipe above or below the ground.

4.4.2 Micro-irrigation system

A micro-irrigation system can be installed on a newly developed landscape with no existing irrigation system, retrofitted on a traditional sprinkler system, or added on to an existing system. Micro-irrigation can be adapted to a variety of residential or commercial landscapes, regardless of topography or plant type. However, it is most effective when used to water trees, shrubs, and other ornamental plants that are spaced widely apart, as opposed to large areas of turfgrass or other plants that cover the entire soil area, where spray sprinkler systems provide the more uniform coverage needed.

Micro-irrigation systems help homes, businesses, and institutions water their landscapes more efficiently. Research indicates that micro-irrigation systems use between 20 to 50 percent less water than conventional spray sprinkler systems. Installing a micro-irrigation system instead of a traditional system can save a typical home more than 25,000 gallons of water per year.

As micro-irrigation provides water directly at the root zone of plants at a lower flow rate, it allows the water to soak into the soil, rather than run off, and applies water only where it is needed. The main benefit of micro-irrigation is efficient water delivery at the root zone, although there are a variety of secondary benefits that make this type of system a preferred choice in many landscape settings. With micro-irrigation, the bare areas of soil or mulch between plants do not receive water, which reduces the chances of weed growth. This results in an aesthetically pleasing and healthy landscape, decreasing the need for added herbicides. Furthermore, pools of standing water, which can attract insects, are reduced, resulting in a decreased need for pesticides.

Such systems will help in protecting local water bodies like streams, lakes, and rivers as they reduce runoff due to high application rates that can sometimes be associated with sprinkler irrigation. As micro-irrigation holds great potential to reduce outdoor water use, proper design, installation, scheduling, and maintenance are essential. As with sprinklers and other forms of irrigation, overwatering with micro-irrigation can occur if efficient practices are not followed.

4.4.3 Design consideration

The design of irrigation system is disposition of various elements of the system in order to maximize its effectiveness in an economical manner. The factors taken that takes into account while designing of micro-irrigation system includes:

- slope
- load losses by friction of water with pipes
- location of water resources. Wells, reservoirs, etc
- possible obstacles such as gas pipes or water, power lines etc

Micro irrigation technician before designing, should know:

- how much water is required by different plant types based on where they are located;
- what types of emission devices should be used to dispense the water;
- how and where the emission devices should be placed

The design of micro irrigation should start with new landscape that is organized by plants different water needs, but the system can still be effective on existing landscapes.

Hydrozoning

Different areas of the landscape have varying irrigation requirements based on plant type, soil type, and sun exposure. To achieve the most efficient application of water, the landscape and associated irrigation system should be divided into hydrozones, or different irrigation zones based on the plant water requirements. To reach this goal, irrigation zones should have plants with similar irrigation needs grouped together in the landscape. For example, turfgrass and shrubs have different irrigation needs and should be in different irrigation zones. This will allow the turfgrass to be independently watered with sprinkler irrigation and the shrubs watered separately with micro-irrigation. Tailoring the amount of water applied to each zone can prevent overwatering areas with different water requirements.

Emission Devices

Fig Different Types of Micro-irrigation Emitters

A micro-irrigation system can include several different types of emission devices, depending on soil type, the landscape's topography, and plant types and their associated water requirements:

- ❖ Drip line emitters are tubes with integrated evenly spaced emitters that discharge at a uniform rate.
- ❖ Multiple outlet emitters have a centralized assembly with multiple emission points that can be connected based on irrigation requirements.
- ❖ Point-source emitters discharge water at a single emission point and can extend from the lateral pipe.
- ❖ Microsprays spread water over a larger area, but still at a low pressure and low flow.

When designing a system where the water supply pressure is greater than that recommended by the manufacturer, emission devices should include pressure compensation to ensure a consistent flow rate despite varying supply pressure, resulting in a more efficient application of water. Emission devices with built-in pressure compensation will maintain a consistent flow rate. Connecting a pressure regulator at the water source will compensate if the incoming water pressure is too high. In-line pressure-regulating valves can be used throughout larger systems with slopes that have intermittent high-pressure areas.

4.4.4 Design factors

Micro-irrigation methods which includes drip, sprinkler and micro sprinklers supply water slowly by depositing water on soil surface or root zone by using network of valves, pipes, tubing and emitters.

Such micro-irrigation methods are efficient water application methods which are adopted to increase the water demands from non agriculture sectors. There exists many considerations which should be considered in order to select an appropriate micro irrigation system such as crop to be grown, water availability and quality, topography, soil type, weather conditions, labour availability, energy, technology and financial resources.

The design of micro irrigation system limited to local climate and crops grown in areas should be developed for good inputs uses. Decision Support System like Design of Micro Irrigation Systems will help the user in designing correct micro irrigation systems that could be used for efficient water utilization under different agro-climatic conditions for any crop under different sources of water availability.

The design of an irrigation system is a complex task due to large number of factors involved in the process. The micro irrigation system design includes design of:

- Drip Irrigation
- Sprinkler Irrigation System
- Micro-Sprinkler Irrigation Systems

Considerations

Water Source Connection and Components

The first step in installing a micro-irrigation system is to identify and connect to the water source, also known as the point of connection. For smaller residential landscapes, the water source may be the outdoor hose bib. For larger systems or systems using well water, the water source may be an existing valve. For retrofit systems, the point of connection may be an existing sprinkler head replaced by a drip zone kit. A filter, pressure regulator, and backflow preventer will likely have to be connected to the water source. An irrigation controller is connected to the valves.

Depending on the source water quality, additional filtration may need to be installed. Water provided by the city should not require any additional treatment beyond a basic filter. However, if the water source is untreated well water or surface water, there may be a need for filtration, such as screen filters, as these sources may contain soil particles or organic matter. Even if the water is relatively clean, using a filter will prevent the buildup of particles over time and extend the life of the irrigation system. Additionally, a backflow preventer may be required by local codes to protect the potable water supplying the house from contamination by the irrigation water.

Many irrigation systems require pressure regulators to maintain a constant pressure, even under varying incoming pressure conditions. The maximum allowable pressure will vary depending on the system design, but should typically be kept at less than 30 psi. Drip line emitters, point-source emitters, and microsprays are designed to operate efficiently between 15 and 30 psi. Manufacturer guides will have specific information on flow rate performance and different incoming pressures for each emission device. Water pressure higher than 30 psi may result in unexpected increases in flow, reducing efficiency and possible system failure. In these cases, a pressure regulator should be considered.

Routing Lateral Lines and Installing Emission Devices

After the water source has been connected and proper filtration and pressure regulation have been installed, the next step is to route the lateral water lines to the appropriate area of the landscape. Where the landscape is divided based on plant type and water requirements, the lateral lines should be positioned to deliver water to each zone. Drip lines and the appropriate emitters should then be installed to supply water to the root zone of the plants. If the area is not properly hydrozoned, it is important to tailor the number and type of emitters to each plant to deliver the appropriate amount of water.

There are adapters that can be used to replace a single sprinkler head with a micro-irrigation line. These can be used in situations where only a few plants need watering and won't require replacing an entire system. After installing the lateral lines, drip lines, and emission devices, check the system to confirm that everything is connected properly. This can be done by turning the system on and inspecting it for leaks, looking for pooling water in undesired places. Examine each emission device to confirm that the flow is consistent throughout the zone. If water is not flowing from an emission device, it may be clogged or blocked, or that area of the system may not be receiving water.

4.4.5 Irrigation Scheduling

There are many factors which a micro irrigation technician has to consider when developing an irrigation schedule which should cover:

- water requirements of plants in each zone
- local climate and weather
- soil composition
- sun exposure
- slope
- depth of root zone

Plant Water Requirement

The plant type and size impact how much water a plant will require to remain healthy in the landscape. Information regarding plant water requirements for common landscape plants is often provided by local utilities or extension offices, as well as online. The age of the landscape also impacts the schedule. Immediately following plant installation, landscapes at their early stages require frequent irrigation intervals to establish the plants. Once the landscape is established, however, the schedule should be adjusted to reduce the irrigation frequency appropriate for those plants.

Weather Conditions

Local climate and variations in evapotranspiration play a significant role in determining how irrigation schedules should shift over the season. In addition, specific weather events, such as rainfall, should be considered. If scheduling manually, take these parameters into consideration and create monthly schedules.

A variety of irrigation controllers and sensors on the market can use local conditions to determine watering times and amounts. Water Sense labeled weather-based irrigation controllers have been independently certified for efficiency and performance; they use local weather and landscape conditions to tailor watering schedules to actual conditions on the site. Similarly, soil moisture-based control technologies use moisture sensed in the soil to adapt scheduling to plant water needs. Additional sensors, such as rain, wind, and freeze sensors, can also help delay or prevent irrigation if conditions are not ideal or necessary for watering.

Soil Types

Soil type should also be considered when developing any irrigation schedule. The infiltration rate of the water depends on the composition of the soil. Coarse soil requires frequent irrigation intervals; medium loamy soil requires longer, less frequent watering; and fine clay soils have a high water-holding capacity that requires the least frequent irrigation. If soil type is not taken into account when developing a schedule, water could be wasted due to runoff or deep percolation. For example, if a coarse soil is not irrigated properly, the water will infiltrate to the deep soil quickly and without enough time for the plants to access it. Irrigation water on a clay soil does not infiltrate quickly, and if the system is run too long, the landscape will begin to flood and runoff could occur. Figure 4 demonstrates how water infiltrates into the various soil types over the same irrigation period.

One way to make irrigation in different soil types more efficient is by including cycle and soak in the irrigation schedule. Rather than irrigating each zone fully in a single cycle, the cycle and soak method irrigates each zone in short intervals and allows water to infiltrate between intervals. This is popular for sprinkler irrigation, but can still be useful in micro-irrigation, depending on the flow rate of the system. If the flow rate is slow enough, there will be no pooling on the surface during the entire cycle and no need to adjust the schedule.

Slope and Sunlight

The slope of the landscape and its orientation to the sun should also be considered when developing an irrigation schedule. Care should be taken so as to not schedule long irrigation events on slopes, as this could result in water running off the landscape. Similarly, ponding is more likely to occur at the lower end of the landscape due to runoff from the elevated areas, so lower-flow-rate emitters should be installed, and/or more frequent, shorter events should be scheduled. Regarding sun exposure, plants exposed to intense sunlight, or on slopes with southern or western exposure, might require more irrigation due to the increased evapotranspiration of those plants.

Depth

Lastly, the depth of the root zone should be considered when scheduling irrigation. If irrigation does not reach the full depth of the root system, plants are not receiving enough water to maximize growth. Both plant type and age should be considered when determining root depth and associated schedule parameters.

4.5 Plan layout for Micro Irrigation System

4.5.1 Survey

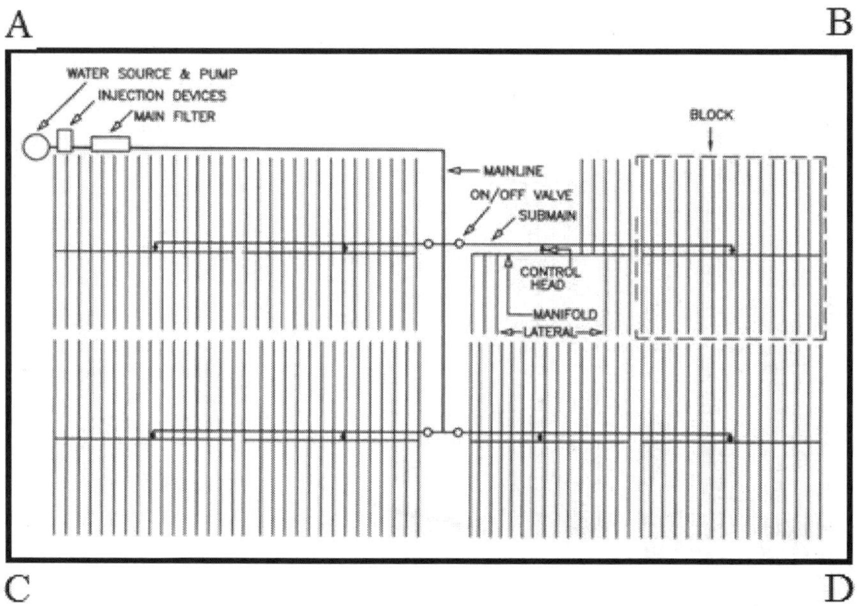

To prepare an accurate layout design of micro irrigation system, the survey inputs are required which should cover:

1. Straight distance between points at corners (AB, BC, CD & DA): It can be measured with a tape in a straight line with corner points duly identified by setting down stones or sticks.

2. Angle at the corner: For a three-cornered area, distances of three of the sides is sufficient to make the layout. For a four-cornered area, any one angle has to be measured along with distances of all sides. For a five-cornered figure, two consecutive angles are required and so on for multiple sides. A distance of 10 meters is marked from the corner on each line, forming the angle, and then a tie length is measured between these points. To determine the corner angle, use the following equation.

$$\tan(\text{angle}) = \text{Length of opposite side} / \text{Length of adjacent side}$$

3. Elevation: Slope of the ground surface may be judged with the naked eye for small plots wherever possible and taken into consideration while designing the drip system. If the ground surface is too undulating and the slope is difficult to judge, levels should be taken

with a leveling instrument and contours drawn on the map to make a proper design of the drip system.

4. Water Source: Position of water source should be marked on the map and the following details noted.

- ❖ Size, volume, flow rate, and height above ground level or depth from ground surface or water source.
- ❖ Pump details for the existing pump including suction, delivery, actual discharge & head, operating time, pump HP, expected discharge & head.
- ❖ Quality of water, impurities in water (algae, sand /silt, etc.) If a water analysis report is available, it should be enclosed with the survey report or if possible the farmer should try to have it analyzed at local laboratory.

5. Agro-climatic details: The details of existing or future cropsshould be noted including specific areas, crop spacing, number of plants and number of rows, crop duration, expected canopy, rainfall, evapo-transpiration, etc.

6. Soil details: The details of soil quality visible to the naked eye should be noted including heavy soil or light soil depending on soil texture. If a soil analysis report is available it should be enclosed with the survey report or the farmer should try to have it analyzed at a local laboratory.

7. Permanent details of the land: Location of farm house, large trees, rocks, etc. should be marked by taking angular measurements from a minimum of two points so that they can be plotted accurately on the survey plan.

4.5.2 Survey Plan

Based on above information, micro irrigation technician has to prepare of survey plan of area surveyed. For smaller areas, an appropriate scale can be used depending on the size of the area. The system layout can be prepared on this plan and then it can be used for installation.

4.5.3 Design Consideration

The micro irrigation design should have both input and output parameters.

Inputs:

The micro irrigation design system kit provides high irrigation efficiency and uniform distribution of water and nutrients for high value crops as compared to conventional flood irrigation systems. In this, when big system is required by farmer, it can be designed in allowable discharge variation limit by using certain procedure. The inputs required to make an effective customized drip micro irrigation system are:

- ❖ Layout of area
- ❖ Details of water source and soil type
- ❖ Agronomic details (plant spacing, crop period, season, canopy, etc.)
- ❖ Climatic data (rainfall, temperature, evapo-transpiration, etc.)

Outputs:

With this, a complete micro irrigation system can be designed using following outputs:

- ❖ Detail layout of system in field
- ❖ Emitter selection and placement
- ❖ Size and length of mainline, sub-main and lateral pipes
- ❖ Pumping and filtration requirement
- ❖ Operating schedule for irrigation.
- ❖ Material and cost estimate

4.5.4 Design Parameters

The design will help the farmer/technician in finding economic feasibility with detailed cost calculation and provides them with default data and guidance for selecting suitable micro-irrigation components.

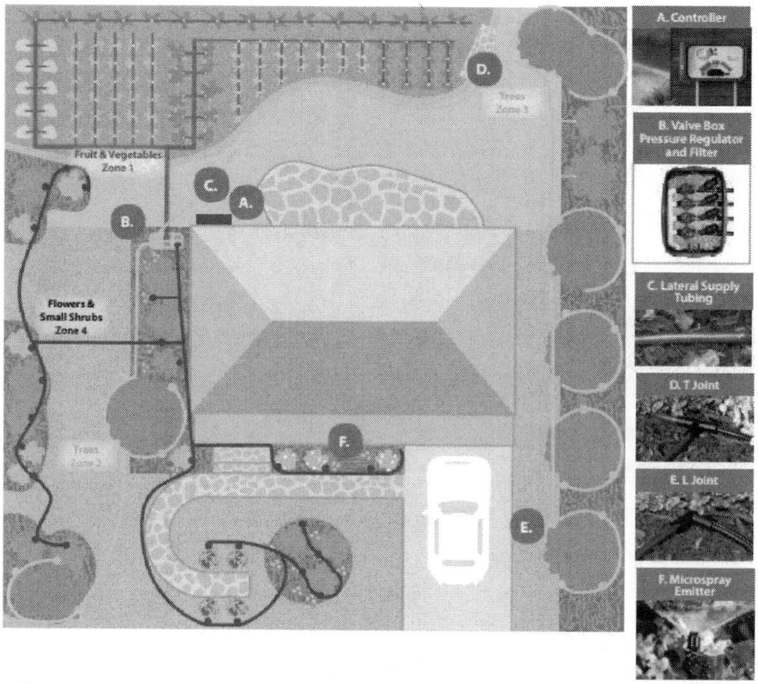

For this, micro irrigation technician or farmer has to follow the design steps, which includes:

❖ partitioning of field into blocks of required dimensions

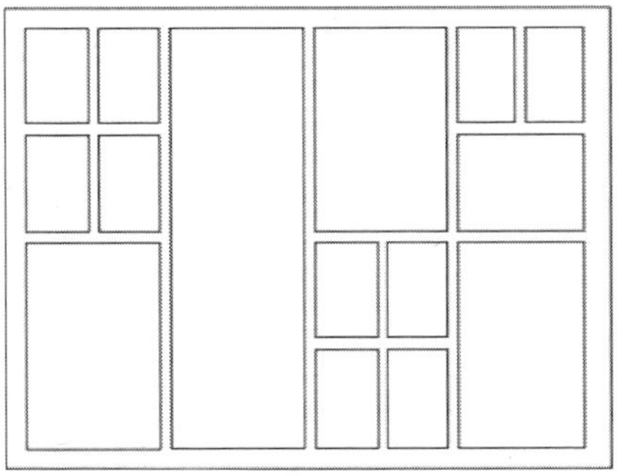

❖ determining appropriate layout plan for main, sub main and lateral pipes in the field

❖ estimating water requirements at plant, lateral, sub main and main pipe level as per agro-climatic data

❖ determining suitable water application rate as per soil properties

❖ determining the sizes of laterals, sub main and main pipes as per their frictional loss analysis

❖ estimating size and type of accessories like filters & fertigation equipment as per water quality and crop demand

❖ determining largest fraction of area which can be irrigated at a time on well source

❖ determining possible shifts and total time for each irrigation

❖ determining total head requirement like lift involved from ground water data

❖ determining size of motor pumping unit and power need

❖ estimating cost of system as per costs of each part

4.5.5 Procedure

❖ System design starts with selection of suitable emitter depending on type of crop, water requirement, operating time, soil type, and water quality.

❖ The length and size of lateral lines are determined based on the lateral line flow rate, field size, etc and also size and length of sub-main pipeline is determined.

❖ Each sub-main is individual unit with its own control valve.

❖ The whole area is then divided into different sub-main units and number of sub-main units that can operate at any one time is based on the existing pumping / water source capacity.

❖ Sections should be designed such that the discharge is similar for all the sections.

❖ The length of the main pipe can be determined based on the flow rate so that frictional head loss is within specified limits and total pressure head required for the system is within pump / water source capacity.

- ❖ If there is no pump, then the pump requirement is worked out from total discharge and pressure head required for the system.
- ❖ Depending on the flow rate and water quality, a suitable filtration device is selected.
- ❖ The total quantity of all the components is calculated from the layout to prepare a cost estimate.

Chapter 5

Installation of Micro Irrigation Systems

5.1 Identify, select, use appropriate tools for Installation

A micro-irrigation system is a small irrigation system which is of two main types:

- drip or trickle systems
- sprinkler systems

Drip systems provide slow dripping of water in particular area while sprinkler systems provide a small jet or spray of water which covers wider area.

Tools

When planning a micro irrigation or drip irrigation installation, it's important to account for special tools and accessories that are often necessary to make the system functionable.

1. I-TOOL

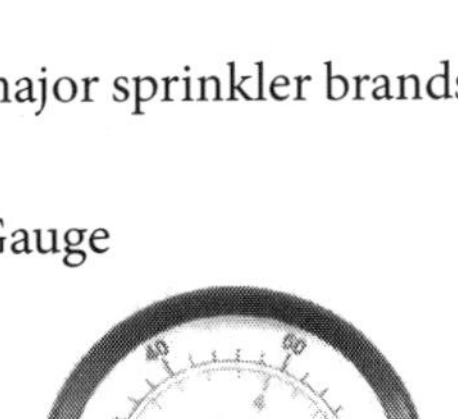

The I-Tool is compatible with all major sprinkler brands and gives you a one stop shop for adjustments, and maintenance.

1. Female Hose Thread Pressure Gauge

2. Netafim Drip Emitter Insertion Tool

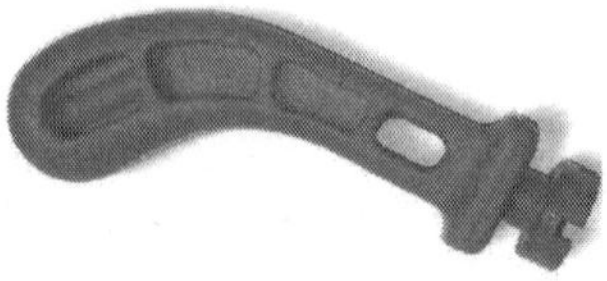

3. Tubing Cutter with Stainless Steel Blade

It is a versatile and lightweight tool that easily cuts many different kinds of thin wall plastic pipe; hose and tubing. In this, spring action is perfect for making multiple cuts fast.

4. Bypass Pruning Shear with Stainless Steel Blades & Alluminum Die-Cast Handles

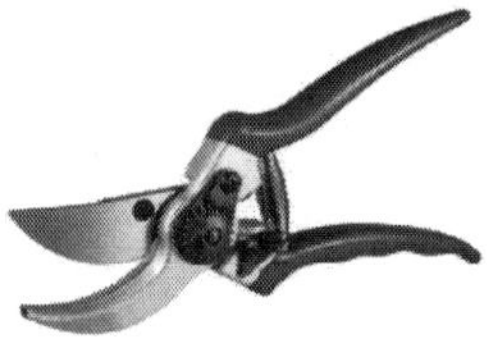

It is a lightweight, strong and sturdy due to its special aluminium alloys. It has awesome cutting power and precision with smooth, tight action with good cutting quality.

5. Replacement Blade for Bypass Pruning Shear

6. Irrigation Wire Stripper & Cutter

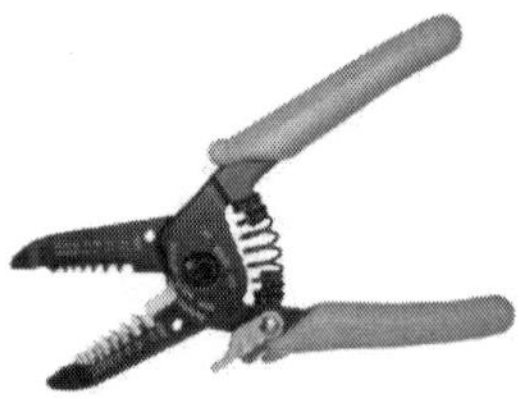

It is of tough steel construction with built-in wire cutter and wire looping holes. The plier nose gives extra gripping strength. In this, laser-etching wire gauge makes spring-loaded automatic opening and gives a black oxide finish.

7. Short Handle Crimping Tool

8. Standard Roll Electrical Tape

9. Contractor Grade Pink Teflon Tape

The contractor grade pink teflon tape is superior in warm conditions and requires less material for each fitting application.

10. Stakes & Risers

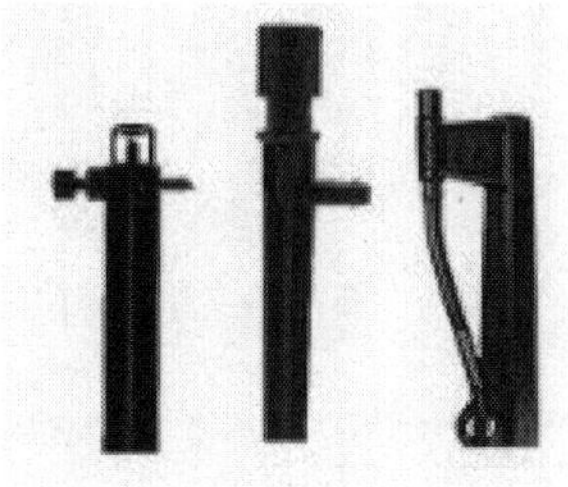

When plants and bushes grow to heights where standard drips and sprays are no longer tall enough to get the water up and over the plants you need a little something extra to help them get up and over.

11. Drip Shrub Adapters

12. Hole Punch

13. Stakes Hold Downs and Clips

It is an important part of irrigation system which makes the installation to look discreet and tidy and ensures intact placement.

14. Hunter Rotor Adjustment Tool

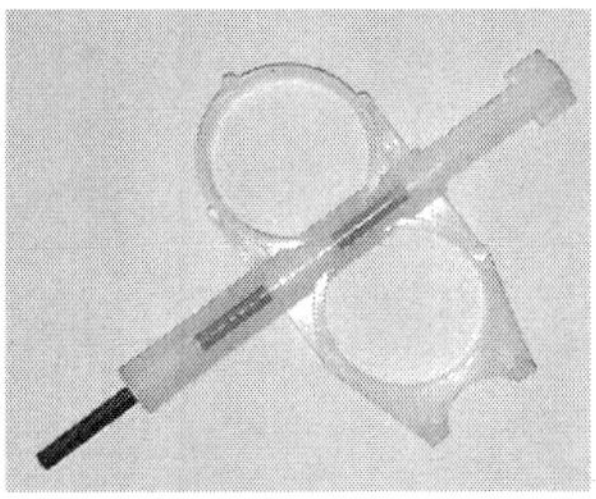

15. Plier Punch

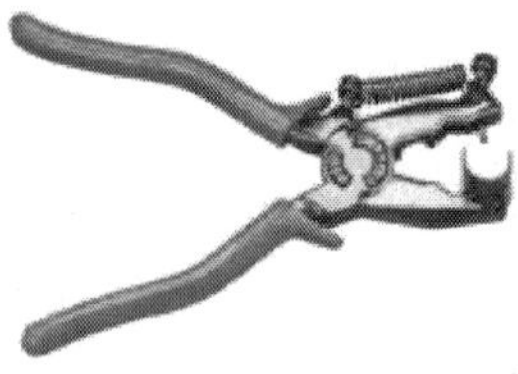

16. Emitter Installation Tool

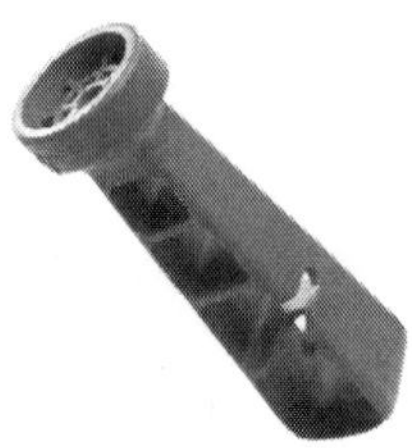

Emitter installation tool works with ¼" barbed couplings and with ¼" barbed tees. It can be used to insert tubing plugs.

5.2 Identify, select and use Micro Irrigation Components for installation

Micro irrigation serves as essential part of efficient farming which helps growers to get maximum yield from their crops for the amount invested. Although there are few varieties of micro irrigation systems with several different options, they all involve basically the same setup. Most micro irrigation systems is fairly simple and can be accomplished by farmer using the parts as discussed below.

From the source of water, to the plants being irrigated, micro drip irrigation systems are comprised of three main parts:

- control system
- delivery system
- emissions system

Control System

The control system handles the entire micro irrigation system that delivers a specified amount of water to plants based on the size of system, its components, and other variables that affect water delivery which comprises of:

- Flow Meters – It controls the amount of water that passes through the micro irrigation system from the source of water.
- Pressure Gauges – It is used in many places throughout the system in order to measure the water pressure in different areas.
- Control Valves – It controls the flow of water to different zones when systems are designed to irrigate multiple areas.
- Irrigation Controllers – These are electronic control valves which are designed to stop and start the system, as it is based on how they are programmed.

Delivery System

The delivery system distributes water from the source to the emitters that irrigate the plants which includes:

- Pipe, Fittings and Tubing – It is used throughout the irrigation system in order to move water from the pumps to irrigation heads. In this, PVC is normally used as mainline delivery to fields, while tubing is used from mainlines to emitters. Also, the fittings in this, connect various sections together. Plastic pipes are flexible black polyethylene pipe having most common sizes as 19mm diameter, used for larger systems as main lines, or more commonly 13mm diameter, for mains and laterals. Pipe with 4mm diameter micro-tubing is used to connect nozzles to main supply lines. When soaked in very hot water, such pipes become pliable and can be pushed over connectors to join to other sections of pipe or for 13 & 19mm pipe, for attach connection to tap. The 13 & 19mm pipes are readily available in rolls of varying sizes, from 20m upwards. The 4mm micro-tubing is readily available in smaller rolls.

- ❖ Connectors/joiners – These are two way, three way, four way, and elbow joints used to connect sections of pipe together. This category also includes small off-take connectors which are pushed into small holes pierced into the main and lateral supply lines, and then rigid sections of 4mm micro-tubing (risers), or flexible micro-tubing are connected to them.
- ❖ Clamps – simple plastic clamps that are placed around the pipe over the connectors/ joiners in 13 & 19mm pipe, and then closed by squeezing them with a pair of pliers or similar tool so that they lock. This helps ensure the pipes don't pull apart due to movement or from water pressure.
- ❖ Flush and Vacuum Relief Valves – These are used to clean out the irrigation system and releases any air trapped in lines which prevents water delivery.
- ❖ Pressure Regulators – These are used to control the water pressure to downstream lines, in order to prevent damage due to high pressure.
- ❖ Chemical Injectors – These are installed in irrigation lines for cleaning of system or in providing plant food in irrigation water.
- ❖ Filters – These are set at the start of delivery line when water is sourced from ponds or other reservoirs to filter out dirt or particles in the water which might plug the system.
- ❖ Outlets – A large variety of usually moulded plastic outlets, which discharge the water either as a spray, a jet, a drip, or bubblers. A hole is simply poked in the pipe wherever you want to discharge water and a nozzle or dripper is pushed or screwed directly into the hole, in the main supply line, or into micro-tubing attached to the main supply line. Some of these outlets are adjustable, so that flow rates, hence for micro-sprays the throw of the water, can be adjusted as required. Spray-nozzles are generally available in quarter, half and full circle spray patterns.

Emissions System

Emissions devices differ based on how water must be delivered. These are connected by flexible tubing and are last in delivery system. They differ in type, pressure, and drip rate, all of which must be considered before deciding on a specific type of irrigator. There are various types of emitters such as:

- ❖ Drip Tape or Drip Line – Tube laid on or in ground which covers regular spaced emitters that seep water along its length.
- ❖ On-line or In-Line Emitters – Tubing with emitters delivering water in specific spots. The only difference in these two types is on-line emitters must be hand-installed into tubing at required intervals, while in-line emitters are factory installed inside tubing at various intervals to save installation time.
- ❖ Micro-sprinklers – Small sprinkler heads that emit water in a circle around the root area of plants.
- ❖ Misters – Emitters that deliver a fine, continuous mist of water to plants.
- ❖ Jets – Emitters that jettison water to land in specific areas to send larger amounts of water at one time.

As micro irrigation systems likely uses similar parts, variables like water pressure, line types, and other necessary parts that depends on type of emitter used and type of plant being irrigated. Micro irrigation systems may seem complex, but their basic design is fairly simple. Farmers/micro irrigation technicians should start with the emitters required, and build to gradually connect them to water pump by considering pressure, filters and different watering zones. One system can be used to irrigate multiple zones in various ways if system is installed to account for any differences in pressure and water delivery requirements.

Fittings

There are a huge range of fittings that are readily available from hardware stores, irrigation suppliers, nurseries or garden centres which includes:

- filters,
- pressure reduction valves,
- in line taps for isolating sections of system,
- controllers (both manual & automatic),
- solenoid valves for automatic systems,
- tap connectors,
- stakes (to support risers)

5.3 Install error free Micro Irrigation System

Micro irrigation system in field/garden is one of the single most important things which a farmer/technician can do. Such irrigation systems provide the right amount of water exactly where it needs saving of money and creating stronger, healthier plants. There are some guidelines/rules which micro irrigation technician needs to follow while installing such irrigation system, to ensure error free working and for best results.

By following the rules/guidelines and applying them across the design and installation of such system, there appears such irrigation system which will provide with many years of usage and worry-free operation. A properly installed micro irrigation system will allow plants to grow healthy and strong and saves money on watering costs. For error free installation of micro irrigation system, the micro irrigation technician has to clearly follow the following things while designing and installing such irrigation system which includes:

Not Enough Emitters

- The most common mistake what a farmer/technican do when designing their system is not including enough drip emitters in their plan.
- With good amount of emitters, plants root systems will get the required water, as more emitters leads to happier and healthier plants with no risk of emitter clogging.
- For smaller plant that only requires one emitter, ensure that you use clogging disk to prevent such risk.

Bad Placement

- Placing drip emitters too close or too far apart also not good, as placing drip emitters evenly will ensure plants of proper amount of water without oversaturated areas.
- For error free system installation, micro irrigation technician should place a drip emitter evenly spaced along the plant line with minimum 6" from base of plant.

Poor or No Filtration

- Filtered water is best for overall health of plants and many micro irrigation systems come with filter to ensure that you have many years of trouble free use.
- Using filter with mesh screen of at least 155 is good if you want to give required protection to small orifices of micro-sprinklers and drippers.

Incorrect Pressure

- Having adequate pressure is good for any micro irrigation system as without it, the system will fail and plants get suffered from lack of water.
- Too many emitters on single line will result in lack of water pressure, which could lead to clogging and inadequate watering.
- For error free installation of micro irrigation system, the technician has to use good pressure regulator to ensure that such system will operate correctly.

Improper Zoning

- By Zoning such type of irrigation systems, micro irrigation technician effectively separates the plants that have different watering needs from others.
- In case of trees, shrubs and dry climate plants; the irrigation technician will required three separate irrigation zones to give water to plants as required without over or under watering.
- Micro irrigation technician should also separate the zones by type of soil and size of plants, as there exists different ways to properly zone the system and knowing the correct way to make all difference to plants.

Wrong Watering Schedule

- There are no set of rules for watering schedules, so micro irrigation technician should have to frame their own way to know when their plants require water and how they can plan for perfect watering scheduling.
- For error free installation of micro irrigation system, the technician should know what the plants need at any given period of time all the year and should properly installed irrigation system to give maximum water to plants.

Ordering Incompatible Thread Types

- Incompatible thread types can cause leaks, loss of pressure and eventual damage to entire system.
- When ordering new irrigation system or parts for existing system, micro irrigation technician should make sure that they get right threads that matches with the threaded parts which they have already installed.

- There are different types of threaded part like hose thread and pipe thread.
- As there are certain parts in micro irrigation system that have different types of threads, where many will use pipe thread or hose thread having option of male or female fitting, so micro irrigation technician should know the basic abbreviations of such threads like PT = Pipe Thread, HT = Hose Thread, FPT = Female Pipe Thread, MPT = Male Pipe Thread, FHT = Female Hose Thread, MHT = Male Hose Thread, NPT = National Pipe Thread, GHT = Garden Hose Thread
- For error free installation of micro irrigation system, the technician should know that without using adapter, different thread types cannot be mixed.
- Since Garden Hose Thread offers watertight seal using a washer and pressure applied against it as National Pipe Thread creates a seal using tightly fitting male and female threads.
- Micro irrigation technician should know that GHT never needs anything more than hand tightening while NPT requires cement, Teflon tape or hand tools to have tight seal.

Not Knowing Outside & Inside Diameter of Tubing

- There are many different sizes of drip tubing and knowing which size is important.
- Tubing is measured by inside and outside diameter of tubing and is very general in designation.
- There are many labels such as ½" poly, ¾" poly, 1" poly and so on which relates to difference in size up to 3 different size variations.
- For error free installation of micro irrigation system, the technician has to keep in mind that not all sizes will match others of same rating, so caution should be executed when ordering.
- If a technician plans to buy all of 1/2" poly tubing and fittings from the depot, then all 1/2" tubing and fittings will be compatible.

5.4 Checking complete Micro Irrigation System after installation

Proper installation of micro irrigation system is important for efficient and trouble free operation. Checking of micro irrigation system should occur on commissioning of the system and at regular intervals thereafter to ensure that the system has been installed as per the design and provides a benchmark to refer to over time.

There are certain tools which will help in testing/checking micro irrigation systems such as:

- pressure gauge which has been tested for accuracy and capable of operating up to 400 kPa
- pitot tube attachment
- watch or stopwatch capable of measuring seconds
- catch cans / collectors
- measuring cylinder

- measuring tape
- marker pegs
- anemometer

Once the micro irrigation system is checked with equipments, it is better to test its lateral lines, drip lines and emission devices, so micro irrigation technician needs to check the system in order to confirm that everything is connected properly and system is in correct use.

For checking the system, micro irrigation technician has to turn the system ON and should inspect the system for leaks, looking for pooling water in undesired places. He should also examine each emission device to confirm that the flow is consistent throughout the zone. If water is not flowing from an emission device, it may be clogged or blocked, or that area of the system may not be receiving water.

While checking the micro irrigation system, the technician should also check for:

Pump Efficiency

- The pump supplies pressure and flow for irrigation system.
- An inefficient pump may have substantially higher running costs and could affect crop yield.
- Pump performance evaluation requires comparison of present pump operation against the manufactured performance curve.

Filter performance

- Filter performance can be measured by finding the pressure differential between entry and discharge points of the filter assembly.
- Filter performance should be done directly after filter flushing and the results compared to manufacturer specifications.

Mainline Evaluation

- ❖ In a correctly designed irrigation system, mainline evaluation should not be required.
- ❖ When a problem is found it is often low pressure associated with larger than expected friction loss.

Valve Evaluation

- ❖ Valves control the rate of flow through a section of pipe and, like filters, range from simple manual units to automated and regulated assemblies.
- ❖ Several checks should be performed to ensure that the valve is performing as desired.

Emitter Evaluation

- ❖ The term emitter refers to a suite of irrigation water discharge devices available in various shapes, sizes and configurations.
- ❖ All components of the irrigation system are designed to ensure that emitters discharge water at a uniform rate across an entire valve area so that all plants receive the same volume of water.

5.5. Guide and trained farmers on installation of Micro Irrigation System

The purpose of micro irrigation technology is to optimise use of irrigation water, reduce wastage, improve productivity and quality while reducing the risk of crop failure in period of water shortages, minimising risk of nutrient leaching and environmental hazards and also enhancing food security and farmers livelihood.

Micro irrigation technician along with team of members can guide and trained farmers by given information based on its installation and setup. While training, they should discuss various merits and demerits of micro irrigation system setup and installation.

While discussing with farmers, micro irrigation technician should discuss certain installation process with farmer and should tell them the necessary details regarding the various parts of the system, their functions and maintenance. For training to farmers, the representative should visit the farmers and discuss them to clear certain doubts, if any during the know how of irrigation system designs and setups as the farmer should know about setup details related to cost, reliability, maintenance, required parts.

Micro irrigation technician along with team of members should discuss with farmers based on following points which includes:

- Reliability: In many instances, micro irrigation system serves as main basis for small farmer to produce food and other products. The reliability of such system provides farmer with maximum assurance and is unforgivable when small farmer invests lot of money to buy system, good seeds and fertilizer.
- Long work life: The small farmer requires a system that will continue operating properly for many years. These systems are used for growing vegetables in 2-3 cycles annually. They are collected after almost every crop, to enable soil preparation for the next crop.
- Resistance to clogging: In many instances the water these farmers use is of poor quality. Water is often transferred in open canals, concentrated in open reservoirs and polluted with foreign matter before reaching the system. At present there are filtration systems and drippers with wide water passages that can ensure water flow through drippers without clogging. Since the small farmer does not have the means or the knowhow to clean clogged drippers, it is most desirable that his system will be with self-cleaning drippers, with internal filtering that will prevent dirt from penetrating the drippers. Filters in the field should be with the largest cleaning surface possible.
- Simple to operate: Such systems is simple to lay out in the field and easy to store after harvesting. They have drippers with uniform water distribution, that are easy to operate and are reliable, as per length of irrigation time or water volume applied with length of time.
- Inexpensive: Micro irrigation technician should discuss the cost analysis of such system as many farmers do not prefer expensive systems. Many farmers receive or can receive grants, subsidies and/or favorable loans, to enable them to purchase such system once in a lifetime. It is impossible to purchase a new system very year or two. Inexpensive systems that are capable of working for a short period are good for rich farmers that can purchase new ones every year; farmers like us require a good system that will last us a lifetime.
- Modular: Systems for use of small farmers need to be modular, so that they can be assembled as a kit, without requiring special planning of field. Micro irrigation technician should discuss regarding the modularity of system as farmers believes that good system should irrigate around 500-1500 m^2.
- Operating instructions: Micro irrigation technician should discuss with farmers regarding the operations of micro irrigation system. They can also make the farmers to prepare and understand operating instructions of the system with explanatory drawings, so that farmers can better understand the system.
- Minimum operating pressure: Micro irrigation technician should discuss regarding the different energy sources requires to pump water to achieve pressure, so farmers should be aware of systems working pressure at which it allows water tank to get filled.

5.6 Gain knowledge about Sprinklers, micro sprinklers and its components

5.6.1 Sprinkler

Sprinkler is a device used to irrigate agricultural crops, lawns, landscapes, golf courses, and other areas. They are also used for cooling and for the control of airborne dust.

Such sprinklers can be used for residential, industrial and agricultural usage. It is useful on uneven land where sufficient water is not available. It is useful for sandy soil. The perpendicular pipes, having rotating nozzles on top, are joined to the main pipeline at regular intervals of time. When water is allowed to flow through the main pipe under pressure with the help of pump it, escapes from the rotating nozzles. It gets sprinkled on the crop.

Sprinkler irrigation system allows application of water under high pressure with the help of a pump. It releases water similar to rainfall through a small diameter nozzle placed in the pipes. Water is distributed through a system of pipes, sprayed into air and irrigates in most of the soil type due to wide range of discharge capacity.

Advantages

- ❖ Eliminates water conveyance channels, thereby reducing conveyance loss.
- ❖ Suitable in all types of soil except heavy clay.
- ❖ Water saving up to 30% - 50 %.
- ❖ Suitable for irrigation where the plant population per unit area is very high.
- ❖ Helps to increase yield.
- ❖ Reduces soil compaction.
- ❖ Mobility of system helps system operation easy.
- ❖ Suitable for undulating land.
- ❖ Saves land as no bunds required.
- ❖ Soluble fertilizers and chemicals use are possible.
- ❖ Provides frost protection & helps in alteration of micro climate.
- ❖ Reduces labour cost.

Components

- ❖ Pumping station or Header Assembly
- ❖ By-pass valve
- ❖ Fertilizer tank
- ❖ Filtration system
- ❖ Pressure gauges
- ❖ Control valves
- ❖ HDPE / PVC Pipes

- QRC Pump connector
- Sprinkler Nozzles
- Service Saddle

Types of Sprinkler	Precipitation Rate(mm per hour)
Low Volume Sprinkler	Less than 13
Medium Volume Sprinkler	13 - 25
Large Volume Sprinkler(Raingun)	Above 25

5.6.2 Micro Sprinkler

Micro Sprinkler is perfectly used where drip irrigation is not advisable which keeps plant foliage moist and when overhead watering is required. They operate at low water pressures but throw out a rotating stream of heavy water droplets, with larger flow rates and over much larger areas than drippers and sprays. Such sprinkler provide low precipitation rates, making them suitable for longer watering time with less run off. This makes them better suited to large flower beds and borders, vegetable patches, nurseries. Inverted micro sprinklers are perfect for green houses.

Micro-sprinklers are emitters known as sprinkler or spray heads which are of several types. In this, emitter operate by throwing water through in air, usually in predetermined patterns. Depending on the water throw patterns, the micro-sprinklers are referred to as mini-sprays, micro-sprays, jets, or spinners. The sprinkler heads are external emitters individually connected to the lateral pipe typically using "spaghetti tubing," which is very small diameter tubing. The sprinkler heads can be mounted on a support stake or connected to the supply pipe. Micro-sprinklers are desirable because fewer sprinkler heads are necessary to cover larger areas. The flow rates of micro-sprinkler emitters vary from 16 lph to 180 lph depending on the orifice size and line pressure

The micro-sprinklers provide irrigation using very fine droplets. They have a rotating deflector, called rotor or dancer, who helps deliver a greater diameter coverage, lower precipitation rate than the diffusers, increased droplet size, and better distribution of water. For each type of micro-sprinkler are several types of rotors. The micro sprinklers are ideal for low volume irrigation in horticultural crops, fruit, flowers, greenhouses, nurseries, frost protection and watering gardens.

5.7 Study different components of Micro Irrigation System and its uses

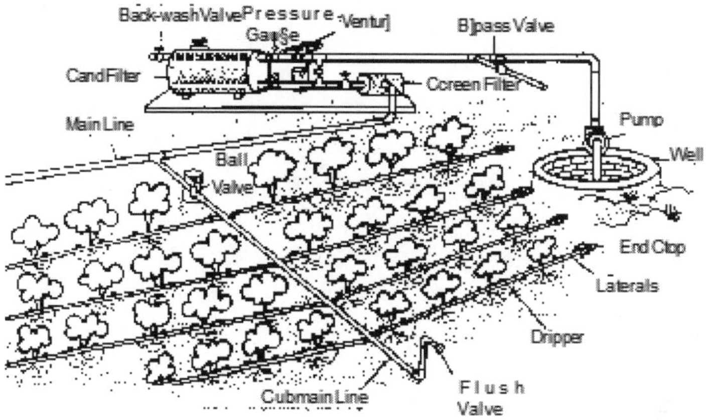

The basic components of micro irrigation system include:

Water Source:

❖ The source of water can be wells, tube wells, canals, river, reservoir etc.

Pump:

❖ The water from the source has to be supplied under appropriate pressure and discharge which is usually done by selecting, right kind of pump.

Sand filter:

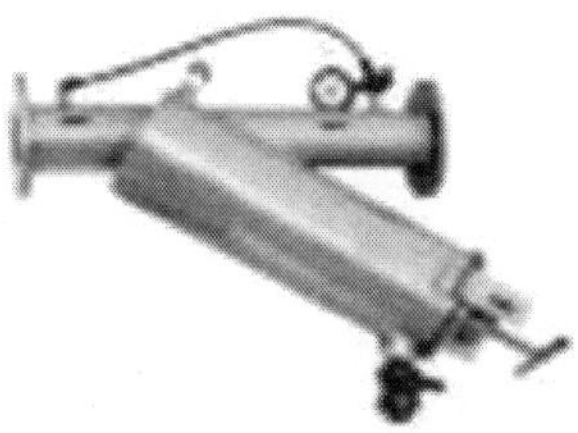
Sand filter

Sand Filter

❖ In micro irrigation system, drippers or drip tape tubes are used which has small holes.

❖ If foreign particles such as leaves, algae etc. coming with water are not filtered, they will clog the dripper and drip tape holes and results in obstruction of water supply to plants.

❖ To prevent such problem, sand filter is connected with main line to stop impurities and allows clean water to go ahead in main line, thus preventing clogging of system.

Screen filter:

Metal Screen Filter

❖ While majority of impurities are filtered by sand filter, minute sand particles and other small impurities pass through sand filter.

❖ Such particles get filtered by screen filter having screen strainer that filters physical impurities and allows only clean water to enter in system.

Pressure Gauge:

Pressure Gauge

❖ Maintaining the normal operating pressure in system is essential to ensure uniformity of irrigation.

❖ Pressure gauge is provided at filtration unit of irrigation system to show pressure which is three way valve arrangements that measures pressure and protect gauge against jerks, pressure surge and vibrations.

Ventury/Fertiliser tank/Fertigation Pump:

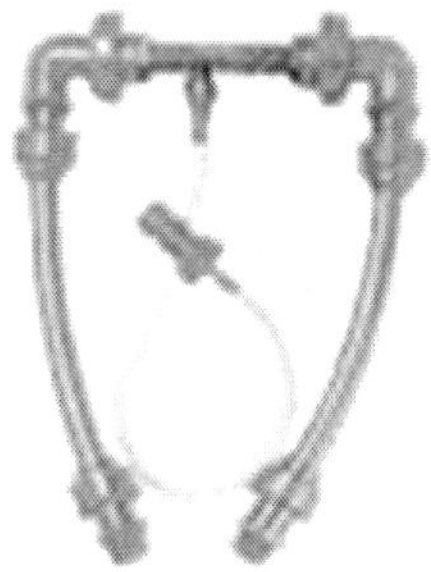

Ventury

❖ Using such equipments, liquid/water soluble fertilizers can be given to root zone directly by injecting such solution with irrigation water.

- It results in full development of roots and crop & prevents wastage of costly fertilizers.
- These are also used in acid treatment and chlorination, to clean drip system & to apply other chemicals.

Valves:

PVC Valves

- Micro irrigation system is provided with different types of valves.
- To control the water pressure, bypass valve is provided before filter.
- Gate valve is provided to create the pressure difference for chemigation and fertigation.
- At inlet of submain, control valve is provided.
- At the end of submain, flush valve is fitted to facilitate regular cleaning of main and submains.
- Air release valve is installed at highest point on mainline to release entrapped air during start of system and to break vacuum during shut off which is provided at inlet of drip tape submain.

Main Line:

PVC Pipes

- In such system, main line is of PVC/POLYETHYLENE pipe lines which are used for carrying water from water source to submain.
- In this, filter is attached with pipe so as to provide filtered water to submain.

Submain Line:

- ❖ Submain is also a PVC/POLYETHYLENE pipe line that supplies water to laterals on one or both sides of it.
- ❖ Submains are buried at 1-1/2 feet in soil to allow cultivation operations in the field.
- ❖ These lines are provided with ball valves to maintain required pressure and flow in the pipe.
- ❖ At its tail end, flush valve is provided for cleaning of main and submains

Poly Fittings:

Poly Fittings

- ❖ Such fittings are used for installing the lateral on submain, connecting the laterals & inter connection between lateral & tape or Turboline, Tee, Elbow, Joiner, Adapter, Take off End Plug & End Stops and are available in 4 - 16 mm size.

Laterals/Drip tape:

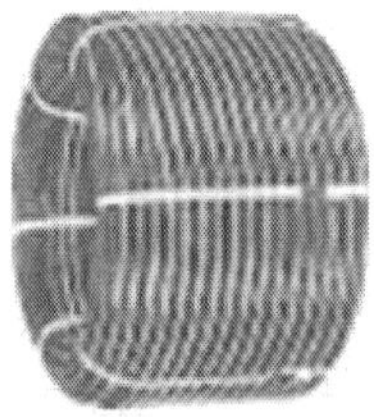

Laterals/Drip tape

- ❖ These are line source tubing or drip laterals which are made of LLDPE which is quite flexible and strong.
- ❖ These convey water from submain lines to root zone via drippers.
- ❖ They spread in open field & their spacing is decided on basis of row to row distance of the crop.

Drippers:

Dripper System

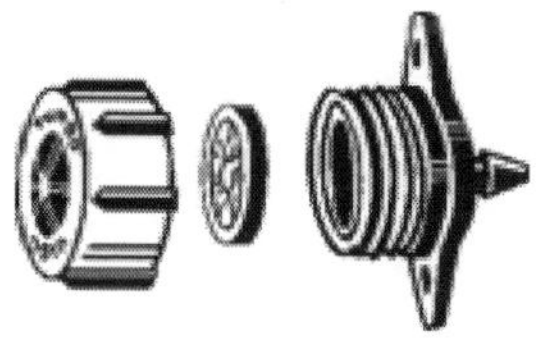

Disc inside the Dripper

- ❖ Water coming from laterals reaches the plant root zone through drippers.
- ❖ Type of crop, soil and crop water requirements are deciding factors for dripper spacing on lateral.
- ❖ Drippers may be pressure compensating or non pressure compensating depending on field topography.
- ❖ Some Drippers are openable to allow cleaning.

Turbo plus J-Flow J-Loc

- ❖ TurboPlus, J-Flow & J-Loc drippers are available with extension barb to connect tube for pot irrigation.

Chapter 6

Maintenance of Micro Irrigation System

6.1 Check and identify the trouble shooting problems occurring in farmer's field

Micro-irrigation is the slow application of water as discrete or continuous drips, tiny streams or miniature spray on, above, or below the soil by surface drip, subsurface drip, bubbler and micro-sprinkler systems. It is applied through emitters connected to a water delivery line through low-pressure delivery.

Micro-irrigation system is complex which requires more filtration and water treatment and has high maintenance costs as compared to other types of irrigation systems. Many problems with micro irrigation system comes from improper installation or by not using right kind of components. There are so many things that can go wrong in a micro irrigation system which can be system inadequately covering the required area, leaving brown patches, uneven water flow, non shutting of systems or part of system which refuse to work.

Micro irrigation system is an arrangement of hoses and tubes that runs throughout the field/garden. In this, the hoses get clogged by debris, salt build up and algae, so if the hoses are covered and hidden by mulch, it becomes difficult to check, in case when they are running properly and fix any clogs. In the system, the hoses are exposed which gets damaged by rabbits, pets, children or gardening tools.

6.1.1 Problems that can be avoided while installing the micro irrigation system

Not installing filter in system

- Many people install the micro irrigation system without a filter.
- Since clean water is required for such system to work, so without filter, it is difficult to clean the normal water.
- Without filter, flow of water will not the proper through emitters which makes the system to jam and stops.
- It also clog the sprinklers, sprayers, emitters, and can even jam the timer.

Ignoring low pressure issues

- ❖ It is easy to overlook checking of pressure at spigot before putting together a micro irrigation system as many people don't have idea about their water pressure and have no time to borrow, rent or purchase pressure meter.
- ❖ If the pressure is too low, pressure-compensating emitters don't work properly and timers and valves won't open.

Not checking the flow

- ❖ Like water pressure, flow of water is also ignored till problem arises.
- ❖ As many municipal systems have 300 GPH of flow, so trouble appears when a micro irrigation system that demands more than source can supply where plants started to suffer.
- ❖ Checking the water flow is simple which doesn't take much time and can save money in future.

Not installing pressure regulator

- ❖ Installing pressure regulator requires space.
- ❖ Many irrigation systems operate best at around 30 PSI, with misters and sprinklers with 40 or 50 PSI.
- ❖ Anything higher than that and fittings can start popping off and mist becomes like smoke and drifts away by breeze making diaphragms in emitters to distort.

Not having extra spares

- ❖ While installing drip irrigation system, technician has to make sure that there should be an extra arrangement of couplers, tees, elbows, goof plugs and other essential fittings.

6.1.2 Troubleshooting problems at site

Problem 1: Leakages at Submain

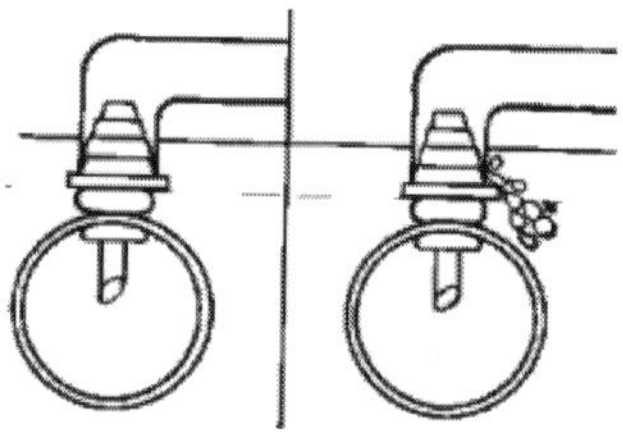

The cause of leakage at Submains are due to:

- ❖ Lateral/polytube at the start is without sufficient allowance.
- ❖ Quite tightly placed polytube with slight pull by animals or people.
- ❖ Grommet takeoff comes out or gets loosened.

Problem 2: Leakages at drip tape/lateral joints

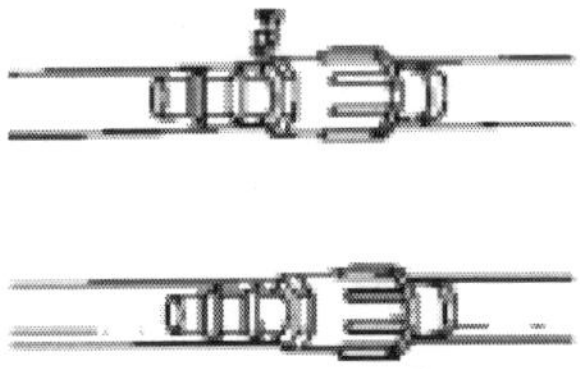

The cause of leakages at drip tape/lateral joints are due to slight jerks, joints get stretched and become loose.

Problem: 3 Leakages along lateral/drip tape

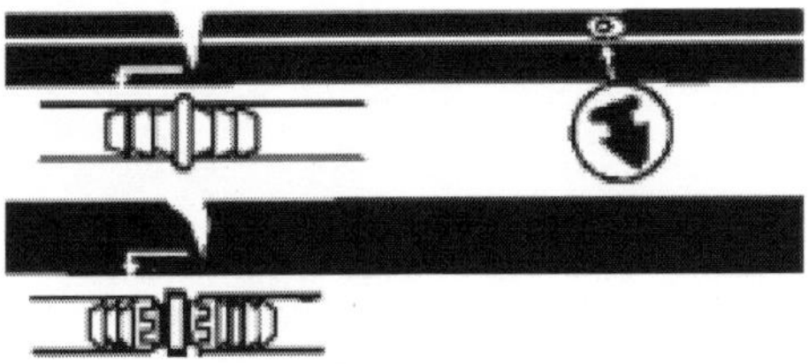

Lateral/Drip tape is damaged by mechanical devices or by squirrels, rats, animals etc.

Problem 4: Difference between inlet/outlet of filter is not normal

The cause of difference among inlet/outlet of filter when more than normal is due to not cleaning of filters or when quantity of sand is more in sand filter.

Problem 5: Pressure gauge doesn't show readings

The cause of pressure gauge not showing readings happens if the gauge is faulty due to rusting or some jerk injuries or when the pointer gets stuck up.

Problem 6: Non-uniform dripper discharge

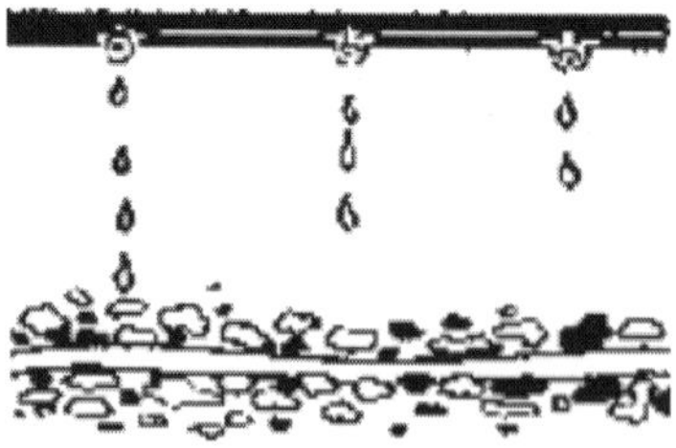

The cause of non-uniform dripper discharge arises when:

- ❖ drippers gets clogged
- ❖ there is an accumulation of algae and salt in drippers and laterals
- ❖ submains and laterals are not flushed properly
- ❖ laterals are pinched or leakages developed somewhere along the length of lateral

Problem 7: Non uniform discharge with dry patches of soil in the drip tape system

- ❖ The cause of non uniform discharge of soil with dry patches in drip tape system happen due to:
- ❖ algae growth in filter or drip tape.
- ❖ blockage of salt accumulation in tape & holes.
- ❖ improper position of air release valve or when it is not working, making silt particles to suck.
- ❖ negative suction, which lead to clogging of holes.

Problem 8: Leakages at dripper placement on the lateral

The cause of leakages at dripper placement on lateral arises when:

- ❖ drippers are taken in and out many times without any use.
- ❖ holes become oblong and bigger in size.

Problem 9: Low discharge with Low pressure reading

The cause of low discharge with low pressure reading is due to:

- ❖ no proper cleaning of sand filters
- ❖ low water level in source
- ❖ improper pump working

Problem 10: No discharge at lateral/drip tape end

The cause of non discharge at lateral/drip tape end is due to cutting/pinching of lateral/ drip tape in between.

Problem 11: Excess pressure at the filters

The cause of extra pressure at filters is due to non bypass assembly.

Problem 12: Ventury is not working

The cause of non working of ventury is due to:

- ❖ lower recommended pressure at pressure gauges.
- ❖ leakages in ventury assembly.
- ❖ improper working of pump.

Problem 13: Presence of sand in screen filter along with trash, straws etc

The cause of presence of sand in screen filter along with trash, straws etc is when sand filter elements are either loosened or not in place.

Problem 14: Dirty water coming from ends of lateral/drip tape

The cause of presence of dirty water at the ends of lateral/drip tape is due to:

- ❖ presence of lot of silt.
- ❖ not proper flushing of laterals/drip tapes for longer time.

6.2 Maintaining the micro irrigation system

6.2.1 Introduction

Micro-irrigation systems include micro-sprinklers for tree crops, drip emitters for trees, vines, row crops and drip tape for row and field crops. Such system applies water to the soil through emitters that are installed along drip lines and contain very small flow passages. The irrigation systems also apply water and fertilizers more uniformly than other irrigation methods. This uniformity results in potentially higher yields, higher revenue and reduced irrigation operating costs.

Uniformity serves as performance characteristic of irrigation systems that measure evenness of applied water all through the irrigation system. It is also called as emission uniformity which serves as an index that describes how evenly or uniformly water is applied throughout the field. A uniformity of 100% shows, same amount of water to be applied everywhere. All irrigation systems applies water at uniformity of less than 100%, and thus some parts of a field receive more water than others. Field evaluation shows that micro-irrigation systems have the potential for higher uniformity than other irrigation methods. However, clogging reduces the uniformity of applied water in micro-irrigation systems, thus increasing the relative differences in applied water throughout a field.

The small flow passages in emitters and micro-sprinklers makes micro-irrigation systems highly susceptible to clogging which reduces uniformity of applied water and decreases amount of applied water. Clogging also decreases the amount of salt leaching around the lateral line in saline soils.

6.2.2 Effect of maintenance

The performance of a micro-irrigation system may rapidly deteriorate if it is not routinely maintained by:

- Checking for leaks.
- Backwashing and cleaning filters.
- Periodic line flushing.
- Chlorinating.
- Acidifying (if necessary).
- Cleaning or replacing plugged emitters.
- Evaluating and monitoring system performance.

Proper maintenance of a micro-irrigation system will:

- Extend system life.
- Improve performance.
- Minimize shut-down time.
- Reduce the probability of non-uniform water and fertilizer applications due to emitter plugging.
- Reduce operating costs.
- Save water and fertilizer.
- Improve production.
- Provide maximum benefits for freeze protection.

6.2.3 Testing of Irrigation Water Source

Before implementing routine or remedial maintenance on micro-irrigation system, test the water source for physical, chemical, and biological properties to determine emitter plugging potential. It is important to take a representative water sample:

- If the water source is a well, collect the sample after the pump has run for about half an hour.
- If sampling surface water, collect the sample near the center of the source about 1 ft below the water surface.

- ❖ Check for water source variability by taking samples several times during the irrigation season.

To accurately measure water pH, alkalinity, dissolved iron, and hydrogen sulfide, analyze well water with a field test kit immediately after sampling.

6.2.4 Routine maintenance of micro-irrigation system components

Routine maintenance involves preventative practices that all micro-irrigation systems should receive regardless of age. Proper attention to following will decrease the likelihood of irrigation system failure:

1. Pumps

- ❖ Follow manufacturer's recommendations to maintain submersed turbine or above-ground centrifugal pumps.
- ❖ Turbine pumps require little maintenance. If failure does occur, repair requires the removal of the pump, which can be complicated and expensive.
- ❖ During the irrigation season, check above-ground pumps at each site visit for:
    - Excessive or unusual noise or vibration.
    - Water leakage.
    - Proper flow rate and pressure.
    - Intake screen obstructions.

2. Power units

- ❖ Electric motor routine maintenance
    - Dirt and corrosion: Wipe, brush, vacuum, or blow accumulated dirt from the frame and air passages. Feel for air discharge from the cooling air ports. If the flow is weak, internal passages are probably clogged and require cleaning. Check for signs of corrosion and repaint or repair if necessary. Open the conduit box and check for deteriorating insulation or corroded terminals.
    - Lubrication: Lubricate bearings only when scheduled, if they are noisy, or if they are running hot. Do not over-lubricate.
    - Heat, noise, and vibration: Feel the motor frame and bearings for excessive heat or vibration. Listen for abnormal noise. Promptly identify and eliminate the source of these problems.
    - Winding insulation: If records indicate a tendency toward periodic winding failures, check the condition of the insulation with an insulation resistance test.
- ❖ Diesel engines
    - During the irrigation season, visually check the engine at each site visit for proper oil pressure, coolant temperature, fluid (oil, fuel, coolant) leaks or stains, excessive noise or vibration.
    - Regularly check the engine oil level with the system off.

- Change engine oil, engine coolant, oil and fuel filters as per manufacturer's recommendation:
- Tune up the engine and take other preventative measures once a year or as the manufacturer recommends.

3. Water filters

Proper water filter performance is critical to minimize emitter plugging. Filters must be periodically cleaned of accumulated particles and debris. Backwashing is a typical cleaning method. A partially clogged filter may reduce system pressure, resulting in reduced and non-uniform water application. Clogged filters also increase pump pressure head and consume extra energy.

❖ Schedule filter backwashing either manually based on a time interval or automatically based on pressure differential.
  - If possible, use automatic backwashing. Set the automatic backwash to operate on a 5 to 6 psi pressure differential.
  - If backwashing manually, determine cleaning frequency based on the length of time it takes for particles to accumulate.

❖ During irrigation periods, inspect screen and disk filters monthly (or more frequently if needed) by removing the cover and examining the filter element:
  - With screen filters, check for tears or extruded material in the screen.
  - With disk filters, check for accumulated organic material on the outside of the disks, and check for sand or other particles that may have become wedged between disks.

❖ Check sand media filters at least twice a year:
  - Check for appropriate sand level.
  - Look for caked material in the media.
  - Make sure media has not flushed out during backwash.
  - Make sure cavities have not opened up.

❖ Routinely inspect all components related to automatic backwashing:
  - Hydraulic tubing.
  - Pressure regulators.
  - Pressure gauges.
  - Control valves.

4. Chemical injection equipment

❖ Visually inspect injection equipment components each time a chemical is injected into the irrigation system:
  - Hoses.
  - Valves.

- Pumps.
- Injector.

❖ Be sure to flush the injection system with water following each chemical injection so corrosive chemicals do not remain in the equipment.

5. Automatic valves

Automatic diaphragm valves are relatively reliable but require periodic inspection to assure proper operation. If a valve failure goes undetected, the pump or power unit could be damaged or water could be applied where it is not needed.

❖ Inspect and clean diaphragm valves at least once a year. A valve can usually be cleaned without removing it from the line.

- Clean deposits that have accumulated on the valve stem.
- Remove encrustation with a wire brush, a weak acid (like vinegar), or very fine sand paper.

❖ When a valve is opened, inspect the diaphragm, seat, and o-ring seals. Replace any components that are beginning to wear out.

❖ Periodically inspect adjustable pressure regulating valves to ensure correct setting.

❖ If regulating valves are pre-set, check them with a pressure gauge mounted at the regulator, or by attaching a portable pressure gauge to a Schrader valve.

6. Pressure gauges and flow meters

❖ Check pressure gauges occasionally to make sure they are working.

- Use high-quality liquid-filled gauges.
- Make sure the range of pressure measured by the gauge covers the operating range of the system.
- Check gauge accuracy by comparing with a new gauge or a standard test gauge.

❖ Occasionally observe flow meters while the irrigation system is operating.

- Make sure the flow rate observed is reasonable for the system.
- Repair or replace a malfunctioning flow meter as soon as possible.

7. Field pipe, tubing, and emitters

Visually check irrigation system field components for leaks each time you visit a running system. Leaks can develop in plastic system parts (often resulting from animal chewing) and in hardware components like pipe fittings, emitters, and hose adapters.

❖ Walk or ride the field, observing or listening for excessive water flow.

❖ When micro-sprinkler stakes are knocked over, the sprinkler pattern becomes grossly distorted. Check for this problem by surveying emitters as they operate.

8. Line flushing

Particulate matter not removed by filters accumulates in irrigation pipes and laterals. Chemical precipitation may occur inside pipelines after the irrigation system shuts down. Suspended materials will be carried with the irrigation water, but as the water velocity decreases near the end of lines, particles will settle. If these sediments are allowed to build up, they will eventually plug emitters.

- ❖ Periodically flush the entire irrigation pipe system (mainlines, submains, headers, manifolds and lateral lines).
- ❖ Manually flush lateral lines by opening only a few at a time. The desired flushing water velocity to remove larger and denser particles is 1 to 2 ft/sec.
- ❖ Determine flushing velocity by measuring the volume of water flowing from an open lateral for 1 minute.
- ❖ Examine the flushed material to get an idea of potential plugging problems. Hold a plastic sheet, nylon sock, or jar at the end of the lateral line to catch the first bit of debris as it leaves the pipe.
- ❖ Determine future flushing frequency based on the amount of material that flushes out:
    - Increase the flushing interval if only a small amount of suspended particles are flushed from the pipe.
    - Reduce the flushing interval if large amounts of material are flushed.
- ❖ After fertilizer is injected, run the system long enough to wash it out of the irrigation system. If chlorine is injected, this extra run time is not necessary.
- ❖ Occasionally irrigation pipes must be cut for maintenance or repairs. Prevent plastic cuttings or shavings from plugging emitters:
    - Use tube cutters rather than saws for repairs.
    - If a saw must be used, clean and flush the repaired section before reconnecting it to the irrigation system.

6.2.5. Remedial Maintenance for Micro-irrigation system affected by Scaling or Plugging

- ❖ Remedial maintenance involves "specialized" procedures that attempt to solve a micro-irrigation encrustation or emitter plugging problem.
- ❖ Preventing emitter plugging is usually more cost-effective than attempting to reclaim a system by chemical treatment.

1. Emitter maintenance and reclamation

- ❖ When an irrigation system becomes severely plugged, replace the emitters with new ones or reclaim the old ones by chemically cleaning them.
- ❖ Before non-plugged emitters are reinstalled, re-analyze the irrigation water to identify the plugging source, and try to identify the material plugging the emitters.

- If the plugged material is calcium carbonate, the chance of reclamation without removing emitters from the field is good.
- If the primary cation is iron, removal and cleaning or replacement with new emitters is likely the only solution.*f* Iron compounds found in plugged emitters are very difficult to unplug in the field. Iron-fouled emitters can be cleaned by soaking in a strong (0.5-1.0%) citric acid solution for 24 to 48 hours.

❖ Reclaiming plugged emitters by chemical treatment is not always successful because most of the injected chemical flows through the open emitters and not through the plugged ones.

- Consider chemical reclamation only as a last resort.
- Direct your major effort towards effective system maintenance.

❖ Do not clean plugged emitters by scraping or reaming with a small wire. Doing so may distort the emitter orifice and can introduce another source of non-uniformity in irrigation water application.

2. Purging with acid

❖ Calcium scale, and to a much lesser extent iron scale, can be purged with acid.

❖ For system purging, lower the pH of the water in the irrigation system to 2.0 or less to achieve maximum effectiveness of the acidification.

- Determine the amount of acid required to decrease the pH to 2.0 or less by titrating a sample of the irrigation water.
- Since water quality can change with time, re-titrate every few months. Use at least 1 gallon of irrigation water, an eyedropper, and a calibrated portable pH meter.
- Stir the irrigation water to ensure complete mixing of the acid before the pH is measured.

❖ Typical acids that may be injected into irrigation water to purge a micro-irrigation system:

- Sulfuric acid – In Florida tests, a 1% solution of sulfuric acid using 38% H2SO4 as the base material removed iron scale from interior tubing walls. However, removal did not necessarily mean dissolution. Sulfuric acid is effective at loosening scale without dissolving it, so line flushing following acid treatment is imperative.
- Citric acid – A 1% citric acid solution removed iron scale from interior tubing walls in Florida tests. In this case, the removal was due to scale dissolution.
- Hydrochloric acid (sold commercially as muratic acid) can effectively remove mineral scale. Hydrochloric acid can be purchased with an inhibitor that minimizes its corrosive effect on metal parts.
- Sulfamic acid is a dry granular material that makes a strong acid when mixed with water. Although it is more expensive than hydrochloric acid and is less aggressive, sulfamic acid offers a number of advantages. In its dry form, it is

relatively safe to handle. Sulfamic acid is particularly useful in treating calcium scale but is less effective on iron.

- Hydroxyacetic (glycolic) acid has been reported to be effective in treating iron scale in wells. Its effectiveness in reclaiming microirrigation system is not known.

❖ Use this general injection procedure:

- Inject acid just long enough for the acid-water mixture to fill the entire interior volume of the irrigation system mainlines, submains, manifolds, and lateral tubing.
- Allow the acid-water mixture to stand inside the system for at least 24 hours.
- Rinse the irrigation system using flush-out valves or by opening lateral tubing ends following acid injection. If the system is not flushed, emitter plugging could be made worse than before if small particles detached by the acid get caught in emitters.

❖ Acids require special filters and injection pump gaskets. Check with the manufacturer of the equipment before acid treatment to ensure compatibility.

6.3 Addressing common problems that occur in the process

After correct installation of micro irrigation system, it is the duty of micro irrigation technician to properly check the system for an errors and to identify any non working part. While checking the micro irrigation system, field technician has to take care about following:

Spouting water at odd angles

❖ If the water is spouting at odd angles or area, then micro irrigation technician should check the sprinklers as there are chances that something is broken.

❖ To fix such issue, field technician has to check for the clogging of dirt at sprinkler head or checking the correct water pressure, as PGP or MP rotators needs different pressures to operate. Also check for matching of sprinkler heads to ensure accurate and uniform distribution of water.

Leakage of water through pipes

❖ If the water is leaking out from the pipes or pipe joints, then micro irrigation technician should check the complete pipe as there are chances of pipe broken or splitting due to ageing.

Plugging of drip hoses and emitters

❖ Solid materials that make their way into drip hoses can cause plugging of drip emitters.

❖ To fix such issue, field technician has to:

- Choose an appropriate filter type which can be media, tubular, disc, etc. as per the water quality and flow requirements.
- Clean the filters and follow maintenance routine for cleaning of disc filters, washing tubular screens and/or replacing sand media at regular intervals.

- Treat irrigation water at pumping plant with sulfuric acid and/or chlorine by continuous injection system to keep soluble materials in solution and prevent scaling and treat agricultural reservoirs to limit algal growth.
- Flush the drip hoses monthly during the growing season and allow hoses to drain for longer durations. Also, adjust the inlet pressures so that a flushing flow rate of 1-2 GPM is achieved at the end of 0.625" ID hose for 5 to 10 minutes per line.

Leakage from hose ends, broken emitters and risers

- ❖ If the water is leaking from hose ends, broken emitters and risers, then micro irrigation technician should check the hose ends, broken emitters and risers.
- ❖ To check such leakages, micro irrigation technician should have a system for identifying and repairing these types of problems.
- ❖ For this, he has to drive the perimeter and every other row of irrigation set at startup and should repair such problems immediately.
- ❖ Irrigators/technician should have such parts and tools to be used in field in order to make repairs immediately.

Water and nutrients not applied evenly over the field

- ❖ If the water and nutrients not applied evenly over the field, then in such case, the field technician should check the pressure.
- ❖ Due to high pressure variability in irrigation set, water and nutrients are not applied evenly over the field.
- ❖ To check such problems, field technician has to remove large pressure losses at pump as more electricity is used which reduces the amount of pressure to work with in the field.
- ❖ The field technician should also make regular pressure observations and should adjust the pressure that regulates the valves so that all valves in the set should have same inlet pressure.
- ❖ He should install the pressure regulators at risers, or directly in drip hoses to reduce high field pressures due to elevation changes.

Problem of Wear and tear of parts

- ❖ The parts that are fixed above the ground will get leak, break or fail from time to time due to environmental factors.
- ❖ To monitor such issues, field technician has to:
    - Check on system components regularly for wear and should replace the damage and/or broken parts.
    - Check for pressure gauges at pump station at the start up and at every irrigation event and should replace the damaged equipment and pipes. He should also check for air vents and should clean them at the start of the season.
    - Replace the drip hoses and emitters when performance is degraded by leaks, cracks and broken emitters.

Seeping of water from sprinkler

- If you see that the water is seeping out of the sprinkler even if the sprinkler is off, then field technician has to check for the solenoids. Such problems appear as due to blocked solenoid valve, so the technician has to check for debris or sediment stuck-up inside the valve.
- Also, if there is pool of water around the valve, then such problem can be due to broken valve or broken pipe, so in such case, field technician has to check both the valve and pipes before replacing any parts.

Excess water supply in fields

❖ Nowadays, every water wise garden needs irrigation timer, that can be manual or automatic which runs by irrigation controller.

❖ If there is excess of water supply in the fields, then irrigation technician has to check for the working of irrigation timers and controllers and should make sure that they set correctly to avoid overwatering.

6.4 System cleaning as per field situation

6.4.1 Introduction

Drip irrigation is considered as the most efficient irrigation system, but the system can also be inefficient due to clogging of the emitters as a result of bad water quality, mismanagement and maintenance problems.

Clogging can be caused by one or more of the following:

❖ Silt (mud from dirty water);

❖ Bacteria and fungi (organic material);

❖ Chemicals (from using fertiliser);

❖ Metals (iron and manganese).

Filtration, flushing and chemical treatment of the irrigation water can be used to prevent the clogging of emitters. Ultrasonic cleaning can be used to complement these and keep drippers functional.

6.4.2 Process of cleaning

1. Flushing the system with water at least 3 times a season can dislodge smaller obstructions and prevent clogs from forming in the first place. If your system is really stopped up, an acid flush may be necessary.

2. Insert plugs in all the emitters. Emitters are holes in the tubing of drip system. The irrigation system comes with plastic black or green plugs that fit inside of the emitter holes. For this, press the plugs into the emitters along the length of the tubing and unscrew the cap at the end of mainline if it is not removed already so that water can flow out of it freely. Now plug the emitters to increase the water pressure which helps in loosen clogs.

3. Turn the micro irrigation system handle to the ON position which allows water to freely flow through the tubing and out of the end of mainline, flushing out clogs and

obstructions. It's normal for the water to be brown initially.

4. Check the water for dirt or bacteria buildup as water should run clear. If the water does not run clear and looks brown, then it shows presence of bacteria buildup in system. In such situation, the system should be flushed with hydrochloric or phosphoric acid.

5. Close the main line and remove the caps from all emitters. For this, screw the plug into the end of the mainline and remove the caps which need to be inserted in emitters. Turn the system back on and examine the emitters that aren't dripping water.

6. Apply muriatic acid on clogged emitters. By using the dropper, drip the acid on the clogged emitters and turn the system ON to see if the acid clears the clogged emitters. If the emitters aren't producing any water, they are clogged.

6.5 Fixing of clogging problems in system

6.5.1 Introduction

Drip irrigation systems provide a favorable environment for bacteria, fungi and algae that can cause slime accumulation. Bacterial slime can be a direct cause for clogging of drippers, but it can also induce mineral particles to stick together and form aggregates large enough to clog the emitter openings. This phenomenon is specifically significant when manganese, sulfide and iron are present in the water.

Depending on the type of bacteria responsible for the slime, its color may be reddish, yellowish or grayish. Clogging due to microorganisms is common when using water with high biological activity, relatively high iron or manganese levels, and hydrogen sulfide compound. Proper chlorination and disinfection procedures are the key to control biological clogging of drip irrigation systems.

The common elements that may clog drip emitters by precipitation and sedimentation are calcium, magnesium, iron and manganese, where calcium carbonate being the most common precipitate. Water that contain high levels of these elements, and have a pH above 7.0, might potentially cause clogging of drip emitters.

Adding fertilizers to source water (fertigation) can potentially cause clogging of drip emitters due to chemical interactions and high mineral concentrations, exceeding their solubility limit. Therefore, it is advised to perform a jar test, or use an appropriate software, to determine if a specific combination of fertilizers may result in precipitation.

Acid injection, to reduce irrigation water pH, can prevent chemical clogging of drip emitters. In this, surface water carry more biological and physical clogging agents, while ground water are usually characterized by higher mineral concentration, posing a chemical clogging hazard.

To avoid clogging of drip emitters, the technician should:

- ❖ Test your source water and irrigation water for the 3 types of clogging causes.
- ❖ Filter large particles or aggregates, such as sand and clay, before letting the water enter the emitters.
- ❖ Eliminate microorganisms by proper disinfection and water treatment.

- Prevent sedimentation of chemicals in the irrigation system by adjusting pH levels, monitoring solubility limits of fertilizers and avoiding chemical interactions between fertilizers.
- Maintain the irrigation lines and emitters by periodical flushing.

Chapter 7

Maintain Health & Safety at Work Place

7.1 Maintaining clean & efficient workplace

7.1.1 Introduction

Cleaning is a process of getting a clean, free, and energetic environment where there is no risk of health hazard. Clean surrounding makes it easier to organize the workplace so that it is easy to access all sort of material which are required.

Cleanliness is an important quality of life and serves as a habit that is often considered just next to godliness. It is a habit which is not related to wealth of a person, but which depicts the qualities of a person.

Cleanliness is nothing that should be forced on a person but should be properly practiced and be made a lifestyle. A clean and well-maintained person will always succeed in all difficulties if he/she applies it in every path he/she walks. It also help to remain fit and healthy, mentally, physically and socially.

Cleanliness is required for health and wellness of an individuals and society. Health and cleanliness relates to each other and influences each other. In order to be in good health, one should practice hygiene. Hygiene is the practice of maintaining good health and preventing diseases through cleanliness. Cleanliness of body, food and environment contribute to the wellbeing of an individual health. It prevents disease as dirt usually dock on infected pathogens which causes disease to human beings when are exposed to them.

Workplace health and safety involves:

- defined roles and responsibilities
- relevant safety information being on-hand
- demonstrated understanding of policies and procedures
- documentation
- training
- regular maintenance
- common sense

7.1.2 Cleanliness at workplace

A clean workplace can be said to as a place where there is all-round protection from germs to health. It is not a place which looks clean but is a place which is actually germs free for the workers and customers. A clean and organized workplace environment will not just be beneficial for worker but also will help in getting more customers. Cleanliness helps in increasing productivity as dirty place and clean mind cannot work together.

Workplaces in many years have realized the importance of clean and healthy surrounding. There are many reasons for which, one should maintain cleanliness and hygiene at workplace. As workplace is where workers spend several hours during their working, so it is advisable that it should remain clean, germs-free and healthy.

In workplace, the same concept applies as productivity is influenced by the environment. A clean environment will increase productivity as people will work with positive mindset that is free of stress. In health, cleanliness is important as hygienic practices contribute to good health of individuals. A dirty environment predisposes people to diseases, which affects their ability to perform activities and it results in less productivity.

7.1.3 Health and Safety

Health and safety is a legal requirement. All small businesses must have an occupational health and safety program to prevent workplace injury and disease. Such programs must meet certain standards, and workers/technicians should exercise the due diligence in taking steps to meet those standards.

Health and safety involves following components:

- Hazard identification and risk control—Determine which hazards are present in the workplace and take steps to eliminate or minimize them.
- Safe work procedures—Describe in writing how to carry out specific tasks safely.
- Orientation, education, training, and supervision—Prepare workers for the job and make sure they continue to work safely. This is particularly important for new and young workers.
- Safety inspections—Identify workplace hazards so that they can be eliminated or controlled.
- Incident investigation—Find out why an accident or injury occurred so the causes can be corrected.
- Health and safety meetings—Provide an opportunity for workers and supervisors to communicate any concerns about health and safety.
- First aid—Determine what level of first aid is required for your workplace, and make sure everyone knows what to do if someone is injured on the job.
- Records and statistics—Maintain documentation to help identify recurring problems, and ensure that hazardous conditions are corrected.

7.1.4 Workplace Risk

The main causes of risk/accidents due to equipments while working at site include:

- transport (running over or vehicle over turnings);
- falling from height (from fragile roofs, trees etc);
- struck by moving or falling objects (bales, trees etc);
- trapped by something collapsing or overturning;
- livestock related accidents and fatalities;
- asphyxiation / drowning.

Micro irrigation technician can prevent most workplace injuries and illnesses if he/ she identify workplace hazards and take steps to control them. Risk control involves eliminating the hazard entirely or, if that is not possible, minimizing the risks as much as possible. Ways to minimize the risks include the following:

- Select appropriate safety features when purchasing or replacing equipment.
- Modify work processes or equipment.
- Develop and implement safe work procedures for hazardous tasks.
- Ensure that workers use appropriate personal protective equipment and follow safe work procedures.

7.1.5 Improving workplace safety

Creating a safe workplace is an ongoing commitment. Employers can make the company a safe place by doing the following or ensuring that it is done:

- Provide regular safety training for supervisors and workers.
- Lead by example and demonstrate safe work habits.
- Regularly check that workers are following company safety procedures.
- Regularly check that workers are correctly using the protective equipment and devices provided.
- Encourage workers to report illnesses and injuries immediately.
- Encourage workers to report potential hazards.
- Keep workplace health and safety programs current.
- Stay aware of workplace hazards and how to handle them.
- Respond promptly to all health and safety concerns.

7.2 Render appropriate emergency procedures

7.2.1 Workplace emergencies

A workplace emergency is an unforeseen situation that threatens employees, customers, or public by disrupting or shutting down the operations or causing physical or environmental damage. To deal with workplace emergencies, micro irrigation technician should focus on emergency response plan which covers designated actions by employers and employees to ensure safety from fire and other emergencies.

7.2.2 Developing Emergency Response Plan

An accident can occur suddenly and without warning. Those who handle or apply chemicals must be prepared to respond to such events. Micro irrigation technician should prepare an Emergency Response Plan which can be quickly and effectively used if an emergency occurs. The plan should give instructions to respond to and deal with all related emergencies like fires, spills, and vandalism. The plan should also be designed to prevent a simple emergency from turning into major disaster and should contain steps to limit the harm to individuals, community and environment.

An emergency response plan must include:

- ❖ preferred method for reporting fires and other emergencies;
- ❖ evacuation policy and procedure;
- ❖ emergency escape procedures and route assignments;
- ❖ names, titles, departments, and telephone numbers of individuals both in and outside of field to contact for extra information or explanation of duties and responsibilities under emergency plan;
- ❖ procedures for workers who remain to perform or shut down critical operations, operate fire extinguishers or perform other essential services that cannot be shut down for every emergency alarm before evacuating;
- ❖ rescue and medical duties for any workers designated in performing them

There emergency plan should be with actions which needs to be carried out by workers /applicator in case of emergency which should have information related to various procedures to be followed in case of accidents and emergencies. There should be ready access to safety equipment which includes personal protective equipment, fire extinguishers and spill response kit.

The first step in making an Emergency Response Plan is to identify any emergency that can happen. The process of making such plan will help in preventing emergencies that could occur any time during the working. Workers may spot and control hazards before they cause harm. The emergency response plans should prepare for:

- ❖ Accidents causing injury
- ❖ Gas and odour release
- ❖ Pesticide spills
- ❖ Fires or explosions
- ❖ Natural disasters (floods, ice storms)
- ❖ Vehicle transport accidents
- ❖ Threats (phone calls)

Ensure that facility and emergency response services are equipped to handle kinds of emergencies common to storage, handling and transport of pesticides, so micro irrigation applicator needs to ensure that all involved in such plan should be aware of what is expected of them in carrying out the plan. Practice with different emergency scenarios will help workers to come to know about the plan. It is good to review the plan before the start of each use season. Copies of the plan should be kept off site and available to emergency responders.

7.2.3 Workplace accidents

Accidents are unplanned and can even happen when proper care is taken. Spills, equipment failures, or fires can occur when handling, moving, storing, or applying pesticides. All who work with irrigation equipments should be prepared to respond to accidents, spills, fires and breakdown.

Accidents are often sudden and without warning. Correct and timely response to an accident with pesticides is crucial. This is often the key to reducing the extent of injury or damage.

Steps can be taken to prepare a response to accidents or emergencies which include:

- Knowing all areas of risk
- Knowing how to respond to different types of accidents
- Taking necessary steps to reduce risk
- Making and using emergency response plans and kits

Procedures after Accident

To protect yourself and your workers during an accident, micro irrigation technician should follow these personal precautions and encourage others to do the same which includes:

- As soon as possible, wash and shower using large amounts of soap and water to remove all toxic chemicals.
- Pay special attention to the scalp, fingernails, all skins folds, ears, nose, and eyes.
- Wash all personal clothing and protective clothing.
- Clean respirators according to equipment directions.
- Discard contaminated leather and canvas shoes, boots, and gloves.
- Watch for early symptoms of pesticide accidents in terms of headache, dizziness, nausea, sweating, or blurred vision as these symptoms shows immediate or after several hours.

7.2.4 Follow emergency procedures to company standard / workplace requirements

Accidents can and do happen, but good planning can reduce their frequency and ensure appropriate and rapid steps are taken to deal with any incident promptly and correctly. All workers/applicators should be familiar with the emergency procedures and trained in how to handle potential incidents.

The employer must develop and implement written operating procedures, consistent with the process safety information, that provide clear instructions for safely conducting activities involved in each covered process. The tasks and procedures related to covered process must be appropriate, clear, consistent, and well communicated to workers. The procedures should address at least the following elements:

- Initial startup;
- Normal operations;
- Temporary operations;
- Emergency shutdown, including the conditions under which emergency shutdown is required, and the assignment of shut down responsibility to qualified operators to ensure that emergency shutdown is executed in a safe and timely manner;
- Emergency operations;
- Normal shutdown;
- Startup following a turnaround, or after emergency shutdown.
- Operating limits:
- Consequences of deviation
- Steps required to correct or avoid deviation.

Safety and health considerations:

- Properties of, and hazards presented by, the chemicals used in the process;
- Precautions necessary to prevent exposure, including engineering controls, administrative controls, and personal protective equipment;
- Control measures to be taken if physical contact or airborne exposure occurs;
- Quality control for raw materials and control of hazardous chemical inventory levels; and
- Any special or unique hazards.
- Safety systems and their functions.

To ensure ready and up-to-date reference, and to form a foundation for needed workers training, operating procedures must be readily accessible to workers who work in or maintain a process. The operating procedures must be reviewed as often as necessary to ensure that they reflect current operating practices, including changes in process chemicals, technology, and equipment, and facilities. To guard against outdated or inaccurate operating procedures, the employer must certify annually that these operating procedures are current and accurate.

The employer/manager must develop and implement safe work practices to provide for the control of hazards during work activities such as lockout/tagout; confined space entry; opening process equipment or piping; and control over entrance into a facility by maintenance, contractor, laboratory, or other support personnel. Such safe work practices must apply both to workers and applicators.

7.2.5 Emergency Procedures

Dealing with suspected poisoning

If pesticide applicator is unwell during or after spraying, then he/she should:

- Stop working;
- Inform someone;
- Seek medical help immediately by either calling a doctor or get someone help for going to hospital;
- Taking product label(s) and any safety data sheet(s) with you;

If someone else is unwell during or after spraying, then he/she should:

- Stop the casualty working and call for medical help immediately;
- If casualty is conscious and mobile, take them away from work area in shelter and then keep them warm and rest until help arrives;
- If casualty is unconscious or not mobile, take suitable precautions to prevent contaminating and then move casualty away to shelter. Check that airway is clear and place casualty in recovery position;
- While waiting for help, remove any contaminated clothing from casualty without contaminating yourself. Make sure the casualty keeps warm;
- Put contaminated clothes aside for later disposal;
- Provide doctor or hospital with product label(s);

Dealing with personal contamination

- Carefully remove any contaminated clothing immediately;
- Wash contaminated skin and hair thoroughly with clean water;
- If eyes are contaminated, flush with clean running water for at least 15 minutes and cover with sterile or clean non-fluffy pad;
- Seek medical help as soon as possible. Take the product label and any safety data sheet and give them to doctor or hospital.

Dealing contamination with bystanders

- Risks to bystanders and local residents should be considered as part of assessment and emergency plan;
- Consider leaving an unsprayed strip adjacent to property of schools/hospitals and respond to reasonable requests from residents;
- If anyone thinks they have been affected by pesticides, advise them to seek medical advice immediately;

Dealing with Fire

- Raise the alarm;
- Call the fire brigade, telling them that pesticides are involved;
- Do not put yourself in danger or breath in any fumes; only tackle a fire if it is safe to do. Use dry powder extinguishers only;
- Provide site plan which shows location of any gas cylinders and chemicals or substances that may be highly flammable, explosive, corrosive, poisonous, oxidising agents, or give off noxious fumes as well as drains and watercourses;

- ❖ For fires in pesticide stores, give officer the list of products and active ingredients.

Dealing with Road Accidents

- ❖ Keep other people and animals away from the area;
- ❖ Call the police, fire brigade and environmental agency. Tell them that pesticides are involved in accident;
- ❖ Attempt to contain any spill to prevent it from reaching drains or watercourses;
- ❖ Provide emergency services with product labels and safety data sheets.

7.3 Reporting danger from time to time to appropriate person

7.3.1 Introduction

All accidents and incidents and near misses arising out of, or in connection with, work including incidents of violence to staff and any work-related to health must be reported to Health and Safety Team promptly. In case of serious injury, an immediate telephone notification is required while other incidents should be reported as per Accident/Incident Report Form. Reporting of Injuries, Diseases and Dangerous Occurrences Regulations set out legal time limits for reporting of certain types of incident to the enforcing authorities.

In many cases it will be appropriate for injured person's immediate supervisor, or a more senior manager, to complete the form after having discussed the incident with those directly involved. Reporting of incidents, accidents should be encouraged as it is the difficult part as culture of blame which has existed for years, staff members may feel they will be victimized if they report incidents. The vital issue is trust as without trust there is no team, and no teamwork. For system to work, workers who report incidents must trust the manager and management to treat them justly and not blame them unfairly or make them a scapegoat.

7.3.2 Features of reporting

The reporting of Injuries, Diseases and Dangerous Occurrences Regulations require workers/employers, or others who control or manage the premises, to report to the relevant enforcing authority and keep records of:

- ❖ work-related deaths
- ❖ work-related accidents which cause certain specified serious injuries to workers, or which result in a worker being incapacitated for more than seven consecutive days
- ❖ cases of industrial diseases listed
- ❖ certain dangerous occurrences
- ❖ injuries to person who is not at work, which are caused by accident at work and which result in person being taken to hospital from site for treatment

7.3.3 Details of Reporting

Reporting and recording are legal requirements. The report tells the enforcing authorities for occupational health and safety about serious incidents and cases of disease, who identifies where and how risks arise and whether they need to be investigated.

Information required on incident reporting form

- Workers name
- Date of birth
- Date and time of incident
- Location of incident
- Brief, factual description of incident
- Name and contact details of any witnesses
- Harm caused, if any
- Action taken at the time
- Name and contact details of the person reporting the incident

7.3.4 Responsibilities

While the manager in line with their structure and bearing in mind the respective responsibilities of working labours, delegate responsibilities. The manager retains accountability for ensuring that all workers are aware of the need to report all accidents, incidents and near misses. In circumstances where a manager of support department is applicable lead, their responsibilities will mirror those set out regarding works, so for purposes of such policy, the management also include all areas in remit of managerial learning.

Manager must establish effective administrative systems within their area of responsibility for internal reporting of incidents, including near-misses which should include:

- Ensure that workers are made aware of need to report incidents and near misses and of local procedures for this which covers incidents that occur away from place.
- Establishing effective administrative procedure for transmission of written reports to be with Health and Safety Team within five days.
- Ensuring that any work-related absences of more than seven days duration, resulting from an accident or work-related ill health are reported to the Health and Safety Team. Individual workers have a duty to be familiar with procedures for reporting of accidents and work-related ill health. Such procedures are covered in worker health and safety induction and should be followed.

7.4 Practice General Safety and first aid

7.4.1 Ensuring safety

Micro irrigation system setup serves as small to medium sized businesses where employees of such small business don't automatically understand their occupational health and safety obligations and doesn't need written occupational health and safety procedures.

The way to prevent injury and ill health at workplace, and to avoid breaching, strict new occupational health and safety regulations for such business is required to have an effective health and safety policy and program which should include clearly defined set of written safety roles and responsibilities.

Safe practices and a safety conscious attitude by both staff and managers is the best protection against workplace accidents. To help you assess the safety status of your business, answer the following questions, highlighting what's missing and determining where you need to improve. By completing this audit at least once a year, a micro irrigation technician can monitor their workplace health and safety performance to ensure they are always improving.

7.4.2 Ways to keep employees/customers safe at workplace

Proper safety is vital to any operation, as day-to-day activities of production or sales can take in focus off things like safety. To fight with combat that, make sure that you should regularly talk about safety practices and procedures with managers and workers. Following are the points which should be managed to handle safety procedure for employees/customers at workplace:

1) Good Housekeeping practices

- Keeping aisles and walkways clear and even/level
- Adequate clearance of exits, breaker boxes, emergency lighting, fire extinguishers, etc.
- In case of emergency evacuation, helps people see and identify safety equipment, prevents accidental slips/trips/falls, and prevents fires

2) Accident investigations

- Identifying root causes
- Implementing corrective actions to prevent re-occurrence
- Gives employer documentation to record the information so it can be reviewed in the future, if needed

3) Employee safety training

- Promotes awareness
- Educates employees on safety with equipment, safety devices, chemical safety, fire prevention, personal protective equipment, etc.

4) Internal safety inspections

- Employee/employer or a team walk through the facility regularly with the purpose of identifying hazards.

5) OSHA required written programs

- Allowing employers to put an organized plan in place
- Writing of safety programs line emergency action plan, Hazard Communication, Respiratory Protection Program, Lockout/Tagout, Confined Space, etc.

7.4.3 Reducing risk of hazards at workplace

There are many controls that employers can use to prevent hazards in the workplace. Firstly though, it is important to complete hazard identification and a risk assessment in consultation with your staff. This will ensure that the right control is chosen for the hazards that are relevant in workplace.

Common controls used in workplaces can be categorised according to the hierarchy of controls:

- Eliminate the hazard - install more power points to avoid cords on floor, widen aisles
- Substitution - resurface floors with 'less hazardous materials'
- Isolation - restrict access to some work areas
- Engineering controls - improve lighting, mark walkways install drainage, use ramps instead of steps
- Administrative controls - ensure good housekeeping - clean up spills immediately, use signs for slippery or wet floors
- Personal Protective Equipment proper footwear

7.4.4 Provide treatment appropriate to injuries with recognized first aid techniques

While working at the site, workers and labours involves in certain type injuries. It is the duty of technician/supervisor to keep/arrange first aid tools at site, so that in case of any emergency, an immediate first aid could be given to the patients using first aid techniques.

First aid kits and attendants

- Make every worker aware of where the first aid kit is located and how to call the first aid attendant if one is required.
- Post signs indicating how to access first aid.
- If a first aid attendant is required, make sure the attendant holds a first aid certificate of the level necessary for that workplace.
- If you require a first aid attendant, train backup attendants. Ensure that enough workers are trained for this responsibility to cover vacations and other absences.

Common site injuries could be bleeding, burns, eye injury, chemicals inside the eyes wounds and breathing etc.

Control of bleeding

- Direct pressure – use your hand(s).
- Elevate (raise) the limb
- Apply a pad and firm bandage.
- If necessary use clean rags or clothing.

Remember

- Always check circulation below the bandage!
- If there is tingling, numbness or blueness loosen the bandage.

Burns

- ❖ Cool the burnt area with cool water for 10-15 minutes.
- ❖ If necessary, cover the burn with a clean dressing or plastic wrap before removing person to medical aid.

Remember

- ❖ Do not burst blisters.
- ❖ Do not remove clothing that is stuck.
- ❖ Do not apply creams.

Eye injuries

Foreign bodies in the eye(s)

- ❖ Wash the eye(s) with eyewash or clean water.
- ❖ If the foreign body is stuck to the eye surface, do not attempt to remove it.
- ❖ Place a covering over the eye(s) and send for, or take, the person to medical aid.

Chemicals in eye(s)

- ❖ Wash the eye(s) with clean cool water for at least 15 minutes.
- ❖ Wash from near the nose outwards and always wash under the upper eyelid.
- ❖ Send for, or take, person to medical aid.

Minor wounds

Abrasions, small cuts

- ❖ Clean the wound with mild antiseptic or soap and water.
- ❖ Cover lightly with clean dressing.

Breathing

If a person is breathing but unconscious, turn them onto their side to prevent airway obstruction (tongue or vomit).

7.4.5 Reporting details of first aid administered as per workplace procedures

First Aid is the initial care provided to a casualty and is usually given by someone 'on the spot'. A person administering first aid is often not a qualified health care professional and is not expected to perform the duties of one. People administering first aid are needed to provide immediate assistance until a qualified health care professional arrives and takes control of the situation or the casualty recovers.

First aid requirements will vary from one workplace to the next, depending on the nature of the work, the type of hazards, the workplace size and location, as well as the number of people at the workplace. These factors must be taken into account when deciding what first aid arrangements need to be provided.

Records of injuries, illnesses, 'near miss' incidents and other information that has already been obtained to assist in controlling risks at the workplace will be useful to make appropriate decisions about first aid.

A record of any first aid treatment given should be kept by the first aider and reported to managers on a regular basis to assist reviewing first aid arrangements. First aid treatment records are subject to requirements under Health Records legislation.

Micro irrigation technician must provide information about first aid to their workers so that they know what to do and who to contact if they are sick or injured. Information should be easy to understand, accessible and should take into account the language and literacy levels of your workers. Information may be given using verbal methods or visual methods.

The information and instruction on first aid should include:

- ❖ location of first aid equipment and facilities
- ❖ names and location of persons trained to administer
- ❖ first aid procedures to be followed when first aid is required

The information and instruction should be provided as part of workers' induction training and when there are any changes, for example in the location of first aid facilities or in the names, locations or contact details of first aiders.

7.4.6 Recording Medical Information

Knowledge of the pesticide involved in an actual or suspected pesticide poisoning can assist medical personnel in their diagnosis and treatment. In the case of a poisoning emergency, provide medical personnel with:

- ❖ The name of the chemical/parts
- ❖ The Registration Number
- ❖ The type and extent of exposure
- ❖ Symptoms observed and their sequence
- ❖ The length of exposure or amount of pesticide ingested
- ❖ The age and weight of the exposed patient
- ❖ The medical history of the exposed patient
- ❖ The first aid performed

Glossary

Acid: A compound that releases H+ ions when dissolved into solution. Compounds such as hydrochloric acid (HCl) or acetic acid (CH_3-COOH) are acids.

Acidification: The introduction of an acid—such as phosphoric, sulfuric, or hydrochloric (muriatic) acid—into an irrigation system. This practice is mostly done in maintenance to improve the effectiveness of chlorination.

Algicide: A substance toxic to algae.

Anti-siphon device: A safety device used to prevent back flow of irrigation water into the water source by back-siphonage.

Application efficiency: The percentage of water applied by an irrigation system and stored in the root zone available for water use.

Application rate: The average rate at which water is applied by an irrigation system. For drip irrigation, rate is expressed as gallons/hour/100 ft or gallons/minute/emitter.

Backflow-prevention device: A device required by Florida law and preventing contaminated water from being sucked back into the water source should a reverse-flow situation occur.

Bactericide: A substance that kills bacteria.

Base: A compound that produces OH- ions when dissolved into solution. Compounds such as potassium hydroxide (KOH) or sodium hydroxide (NaOH) are bases.

Best Management Practices (BMP): A set of cultural practices known to increase the efficiency of the irrigation and fertilization program while minimizing the environmental impact of production.

Certifying agency: An independent, accredited third party that verifies that a certified-organic operation is compliant with the regulations described in the National Organic Standards as appropriate for their farming system.

Chelate: A compound that binds polyvalent metals at two or more cation-exchange sites. Chelate is often a component of ready-to-use formulations for drip-irrigation cleaning. The use of synthetic chelates is not allowed in certified-organic production for cleaning drip-irrigation systems, but synthetic chelates may be used in certified-organic production to correct a documented micronutrient deficiency.

Chemigation: A general term referring to the application of water-soluble chemicals into the drip-irrigation system. Chemigation includes (when allowable) the application of fertilizers, acids, chlorine and pesticides.

Chlorination: The introduction of chlorine—at a calculated rate—into an irrigation system. Chlorination can use liquid sodium hypochlorite (household bleach) or chlorine gas. Some chlorinating agents are allowed in certified organic production.

Cleaning agent: A substance used to remove dirt, filth, and contaminants.

Control valve: A device used to control the flow of water. Control valves turn on and off water to the individual zones.

Detergent: A synthetic substance that is not a soap and that is used to change the surface tension to remove oil and grease and other substances relatively insoluble in water. Detergents are not allowed in certified organic production.

Disk filter: A stack of round, grooved disks used to filter water in a drip-irrigation system. As the size of the groves decreases, the more the water is filtered. Each disk has groves on both sides. Sediments and organic matter accumulate on the disks as water passes through the groves. Disks are reusable. Once taken apart, they can be easily cleaned with pressured water and/or a detergent solution.

Drip irrigation: A method of irrigation using the slow application of water under low pressure through tube openings or attached devices just above, at or below the soil surface.

Electronic Database Information System (EDIS): The online database where the science-based, peer-reviewed, and up-to-date recommendations of UF/IFAS Extension are accessible (http://edis.ifas.ufl.edu).

Emitter: A dispensing device or opening in a micro-irrigation tube that regulates water application. An emitter creates a controlled flow expressed in gallons/minute/emitter or gallons/100 ft/hr.

Emitter spacing: Distance between two consecutive emitters. Typical emitter spacings for vegetable crops are 4, 8, and 12 inches.

Evapotranspiration (ET): The combined losses of water by evaporation from the soil and transpiration from the plant.

Fertigation: The application of soluble fertilizer (plant nutrients) through a drip irrigation system. Fertigation is allowed in certified-organic systems provided the fertilizer sources used are allowed by NOP standards.

Field capacity: The water content of the soil after all free water has been allowed to drain by gravity.

Filter: A canister device containing a screen or a series of disks of a specified mesh or filled with a coarse solid medium and designed to catch solid particles large enough to clog emitters.

Fittings: The array of coupling and closure devices used to construct a drip system and including connectors, tees, elbows, goof plugs, and end caps. Fittings may be of several types, including compression, barbed, or locking (spin or ring).

Flow: The amount of water that moves through pipes in a given period of time. For micro-irrigation (drip irrigation), flow is expressed in gallons-per-hour (gph) or gallons-per-minute (gpm).

Flow meter: A device used to measure changes in flow in a drip-irrigation system over the course of a crop cycle.

Goof plug: An insertable cap used to plug holes in mainline and microtubes where drip devices have been removed or are no longer needed or when an accidental hole needs to be plugged.

Hole punch: A device that makes round holes in the pipes so to connect drip tape with laterals (available in different diameters).

Hydrochloric acid (HCl): An acid often used to lower the pH of water to increase the efficiency of chlorination. Use of HCI is prohibited in certified organic production.

Hypochlorous acid (HOCl): The weak acid generated by chlorinating products. Hypochlorous acid destroys organic matter. Use of HOCI is restricted in organic production.

Irrigation schedule: The watering plan and procedures that determine the proper amount of water to apply, the operating time, and the frequency of an irrigation event.

Mainline: The tubing used in the drip system. Mainline is sometimes called lateral line. It may be made of hard PVC or soft polyethylene material and comes in diameters ranging from 0.5 to 4 inchs.

Mazzei injector: Patented T-shaped, venturi-type injector that does not involve moving parts.

Media filter: A pressurized tank filled with fine gravel and sand. The sand is placed on top of the gravel. Sharp-edged sand or crush rock are more efficient in catching soft algal tissue than round particles. Media filters should be used for filtering water that contains high levels of organic matter.

Micro-irrigation: Synonym for drip irrigation.

Muriatic acid: Another name for hydrochloric acid (HCl).

National Organic Program (NOP): Federal program created as a result of the Organic Foods Production Act of 1990 and operated under the USDA Agriculture Research Service.

Organic Materials Review Institute (OMRI): A national, nonprofit organization that determines which input products are compliant with the National Organic Program Standards (http//www.omri.org). The use of OMRI-listed products requires the approval of a certifying agency. Listing of a product on OMRI lists is not a guarantee of efficacy.

Overfertigating: Applying more fertilizer than the recommended rate. Overfertigating may result in nutrient leaching below the root zone.

Overwatering: Applying more water than necessary to meet the crop needs and/or applying water in excess of soil water-holding capacity. Overwatering potentially results in nutrient leaching below the root zone.

Part-per-million (ppm): The ratio of one in one million: 1 ppm = 1/1,000,000. The "ppm" measurement may also represent concentrations: 1 ppm = 1 mg/L.; 1% = 10,000 ppm.

Peracetic (acid also known as peroxyacetic): A mixture of acetic acid (CH_3- COOH) and hydrogen peroxide (H2O2) in an aqueous solution that can be used in certified organic production as a substitute for prohibited chlorination products.

Permanent wilting point: The water content of the soil in the plant root zone when the plant can no longer extract water from the soil.

pH: A number between 0 and 14 that represents the amount of acidity (H+ ion concentration) in solution and calculated as: pH = -log[H+]. pH can be simply measured with a pH-meter. A solution is acidic when pH<7, neutral when pH = 7, and basic when pH>7. pH affects the solubility and ionic forms in solution. pH is the single most important chemical parameter for water or soil.

Phosphoric acid (H3PO4): An acid often used to lower the pH of water so to increase the efficiency of chlorination. Use of phosphoric acid is prohibited in certified organic production.

Pressure: The "force" propelling water through pipes. Common static (nonflowing) pressure in irrigation systems is 20–70 psi. Irrigation systems operate under dynamic (flowing) water pressure, which is reduced with elevation gain and friction loss caused by the water rubbing on the sides of pipes.

Pressure due to gravity (in pounds-per-square-inches or psi): This measurement may be calculated as gain (downhill) or loss (uphill) by multiplying the height of the water column in feet by 0.433. For example, if a 200-foot drip tape is laid on a field with a downhill slope of 3 ft/100 ft (3%), the gain in pressure due to gravity will be 200 x 0.03 x 0.433 = 2.59 psi.

Pressure loss: The loss of water pressure under flow conditions caused by debris in a filter, friction in pipes and parts, and elevation changes.

Pressure rating: The maximum pressure a pipe or drip-system component is able to handle without failing. For example, Class 160 PVC pipe refers to plastic irrigation pipe with a pressure rating of 160 pounds per square inch (psi). Aluminum irrigation pipe has a pressure rating of 145–150 psi. These pressure ratings will normally be adequate for mainlines in drip-irrigation systems.

Pressure-relief valve: A valve that opens and discharges to the atmosphere to relieve the high pressure condition when pressure in a pipeline exceeds a pre-set point.

Pressure-compensating emitter: An emitter designed to maintain a constant output (flow) over a wide range of operating pressures and elevations.

Pressure-sensitive emitter: An emitter that releases more water at the higher pressures and less at lower pressures, which are common with long mainlines or terrain changes.

Pressure regulator: A device that reduces incoming water pressure for low pressure drip systems. Typical household water pressure is up to 50–60 psi while drip systems are designed to operate so not to exceed 8–12 psi in the drip tape. Due to friction losses, pressure in the delivery pipes may be 20–30 psi, thereby requiring a pressure regulator.

The important ratings of a pressure regulator are the diameter, the downstream pressure and the maximum flow allowed by the pressure regulator.

Root zone: The depth and width of soil profile occupied by the roots of the plants being irrigated.

Sand separator: A device also called hydrocyclone that utilizes centrifugal force to separate sand and other heavy particles out of water. It is not a true filter, but could be considered a pre-filter.

Screen filter: A type of filter using a rigid screen to separate sand and other particulates out of irrigation water.

Self-flushing end cap: A spring-loaded device that lets water go out at the end of the drip tape when the water pressure is less than the threshold of the cap.

Sulfuric acid (H_2SO_4): An acid often used to lower the pH of water so to increase the efficiency of chlorination. Use of sulfuric acid is prohibited in certified organic production.

Sulfur, powdered: Elemental sulfur (S) in the yellow powder form is allowed in certified organic production. It is commonly used to decrease soil pH, but this requires chemical conversion by soil microorganisms. Powdered sulfur should not be used for chlorination purposes.

Soap: Alkaline salts of fatty acids used to remove hydrophilic particles.

Strong acid: An acid that is totally dissociated in water. Common strong acids are hydrochloric acid (HCl) and sulfuric acid (H_2SO_4).

Tape-to-lateral connector: A device sometimes called a barbed adapter and that is placed at the end of the drip tape to connect it with the lateral (snap end).

Tape-to-tape connector: A device used to repair or replace a leaking section of drip tape. The tape-to-tape connector allows two pieces of drip tape to be connected together.

Trickle irrigation: Synonym for drip irrigation.

Turbulent-flow emitter: Emitters with a series of channels that force water to flow faster, thereby preventing particles from settling out and plugging the emitter.

Uniformity of water application: A measure of the spatial variability of water applied or stored in an irrigated field down a row and across several rows. Uniformity of water application is usually expressed as a percentage, 100% representing perfect uniformity.

Venturi injector: A tapered constriction which operates on the principle that a pressure drop accompanies the change in velocity of the water as it passes through the constriction. The pressure drop through a venturi must be sufficient to create a negative pressure (vacuum), relative to atmospheric pressure. Under these conditions, fluid from the tank will flow into the injector.

Water applied: The amount of water actually applied during an irrigation cycle. For drip irrigation, it is expressed in gallons/100 feet of drip tape.

Water hardness: The sum of multi-valent ions—such as calcium, magnesium, aluminum, or iron—in solution. Hardness is expressed in mg/L of calcium carbonate equivalent, and its value is used to classify the water as soft (0-20 mg/L), moderately soft (20-40 mg/L), slightly hard (40-60 mg/L), moderately hard (60-80 mg/L), hard (80-120 mg/L), or very hard (>120 mg/L).

Water alkalinity: Ability of water to neutralize acids. Water alkalinity is based on the content of hydroxide (OH^-), carbonate (CO_3^{--}) and bicarbonate (HCO_3^-) ions.

Water velocity: The speed at which water travels inside a pipe, usually expressed in feet/second.

Water hammer: Pressure surge that occurs because of sudden stoppage or reduction in flow or because of a change in direction of flow. Water hammer may be reduced by slowly turning water on and also by an irrigation-system design in which water velocity is less than 5 feet/second.

Weak acid: An acid that is only partially dissociated in water. Common weak acids are phosphoric acid (H_3PO_4), boric acid (H_3BO_3), acetic acid (CH_3-COOH), and citric acid [COOH-CH2-(COOH-C-OH)-CH2-COOH].

Zone: A section of an irrigation system that can be operated at one time by means of a single control valve.

References

Clark, G.A. and A.G. Smajstrla. 2006. Treating Irrigation Systems with Chlorine. CIR1039. Gainesville: University of Florida Institute of Food and Agricultural Sciences. http://ufdc.ufl.edu//IR00000971/00001

Clark, G.A., C.D. Stanley, and A.G. Smajstrla. 2002a. Microirrigation on Mulched Bed Systems: Components, system capacities, and management. BUL245. Gainesville: University of Florida Institute of Food and Agricultural Sciences. http://edis.ifas.ufl.edu/ae042

Clark, G.A., C.D. Stanley, F.Z. Zazueta, and E.E. Albregts. 2002b. Farm Ponds in Florida Irrigation Systems. BUL257. Gainesville: University of Florida Institute of Food and Agricultural Sciences. http://edis.ifas.ufl.edu/ae143

Clark, G.A., D.Z. Haman, and F.S. Zazueta. 2005. Injection of Chemicals into Irrigation Systems: Rates, volumes and injection periods. BUL250. Gainesville: University of Florida Institute of Food and Agricultural Sciences. http://edis.ifas.ufl.edu/ae116

Dittmar, P.J., E.J. McAvoy, M. Ozores-Hampton, R.N. Raid, H.A. Smith, B. Wells, J. Beuzelin, J.W. Noling, L. Zotarelli, S. Zhang, C.F. Miller, and Q. Wang. 2017. "Chapter 15. Root Crop Production in Florida." Vegetable Production Handbook. HS965. Gainesville: University of Florida Institute of Food and Agricultural Sciences. http://edis.ifas.ufl.edu/cv300

Florida Department of Agriculture and Consumer Services. 2015. "Water quality/quantity best management practices for Florida vegetable and agronomic crops." Florida Department of Agriculture and Consumer Services, Office of Agricultural Water Policy. https://www.freshfromflorida.com/content/download/77230/2220421/vegAgCropBMP-loRes.pdf

Ferguson, J. 2004a. Definition of terms used in the National Organic Program. HS963/HS209. Gainesville: University of Florida Institute of Food and Agricultural Sciences. http://ufdc.ufl.edu/IR00000275/00001

Ferguson, J. 2004b. General Guidelines for Organic Crop Production. HS972/HS212. Gainesville: University of Florida Institute of Food and Agricultural Sciences. http://ufdc.ufl.edu/IR00000276/00001

Freeman, J.H., E.J. McAvoy, N.S. Boyd, M. Ozores-Hampton, M. Paret, Q. Wang, C.F. Miller, J.W. Noling, and X. Martini. 2017. "Chapter 7. Cucurbit Production." Vegetable Production Handbook. HS725. Gainesville: University of Florida Institute of Food and Agricultural Sciences. http://edis.ifas.ufl.edu/cv123

Freeman, J.H., E.J. McAvoy, N.S. Boyd, R. Kinessary, M. Ozores-Hampton, H.A. Smith, J.W. Noling, and G.E. Vallad. 2017. "Chapter 18. Tomato Production." Vegetable Production Handbook. HS739. Gainesville: University of Florida Institute of Food and Agricultural Sciences. http://edis.ifas.ufl.edu/cv137

Gazula, A., E. Simonne, and B. Boman. 2007. Guidelines for enrolling in Florida's BMP program for vegetable crops. HS1114/HS367. Gainesville: University of Florida Institute of Food and Agricultural Sciences. http://ufdc.ufl.edu/IR00000492/00001

Haman, D.Z. 2017. Causes and Prevention of Emitter Plugging in Microirrigation Systems. BUL258. Gainesville: University of Florida Institute of Food and Agricultural Sciences. http://edis.ifas.ufl.edu/ae032

Haman, D.Z. and F.T. Izuno. 2003. Principles of Micro Irrigation. AE70/WI007. Gainesville: University of Florida Institute of Food and Agricultural Sciences.

Haman, D.Z. and A.G. Smajstrla. 2003. Design Tips for Drip Irrigation of Vegetables. AE260/AE093. Gainesville: University of Florida Institute of Food and Agricultural Sciences. http://ufdc.ufl.edu/l/IR00003290/00001

Haman, D.Z., J.C. Capece, and A.G. Smajstrla. 1997. Irrigating with High Salinity Water. BUL322/AE091. Gainesville: University of Florida Institute of Food and Agricultural Sciences. http://ufdc.ufl.edu/l/IR00003289/00001

Haman, D.Z., A.G. Smajstrla, and F.S. Zazueta. 2003a. Media Filters for Trickle Irrigation in Florida. AE57. Gainesville: University of Florida Institute of Food and Agricultural Sciences. http://edis.ifas.ufl.edu/wi008

Haman, D.Z., A.G. Smajstrla, and F.S. Zazueta. 2003b. Screen Filters in Trickle Irrigation Systems. AE61. Gainesville: University of Florida Institute of Food and Agricultural Sciences. http://edis.ifas.ufl.edu/wi009

Haman, D.Z., A.G. Smajstrla, and F.S. Zazueta. 2003c. Chemical Injection Method for Irrigation. CIR864. Gainesville: University of Florida Institute of Food and Agricultural Sciences. http://edis.ifas.ufl.edu/wi004

Haman, D.Z., A.G. Smajstrla, and F.S. Zazueta. 2003d. Fittings and Connections for Flexible Polyethylene Pipe Used in Microirrigation Systems. AE69/WI011. Gainesville: University of Florida Institute of Food and Agricultural Sciences.

Haman, D.Z., A.G. Smajstrla, and G.A. Clark. 2003e. Water Wells for Florida Irrigation Systems. CIR803. Gainesville: University of Florida Institute of Food and Agricultural Sciences. http://edis.ifas.ufl.edu/wi002

Kidder, G. and E.A. Hanlon. 2003. Neutralizing excess bicarbonates from irrigation water. SL142. Gainesville: University of Florida Institute of Food and Agricultural Sciences.

Liu, G., C.F. Miller, B. Wells, Y. Li, and Q. Wang. 2017. "Chapter 5. Asian Vegetable Production." Vegetable Production Handbook. CV301. Gainesville: University of Florida Institute of Food and Agricultural Sciences. http://edis.ifas.ufl.edu/cv301

Liu, G.D., E. H. Simonne, and G.J. Hochmuth. 2015. Soil and Fertilizer Management for Vegetable Production in Florida. HS711. Gainesville: University of Florida Institute of Food and Agricultural Sciences. http://edis.ifas.ufl.edu/cv101

Liu, G.D., E. H. Simonne, K.T. Morgan, G.J. Hochmuth, M. Ozores-Hampton, and S. Agehara. 2017. "Chapter 2. Fertilizer management for vegetable production in Florida." Vegetable Production Handbook. CV296. Gainesville: University of Florida Institute of Food and Agricultural Sciences. http://edis.ifas.ufl.edu/cv296

Mahovic, M., J. Brecht, S. Sargent, M. Ritenour, K. Schneider, A. Simonne, J. Bartz. 2002a. Fresh Produce Handling, Sanitation and Safety Measures: Beans, cucumbers, eggplants, squash, peppers, sweetcorn. FSHN0213/FS094. Gainesville: University of Florida Institute of Food and Agricultural Sciences. http://ufdc.ufl.edu/l/IR00002058/00001

Mahovic, M., J. Brecht, S. Sargent, M. Ritenour, K. Schneider, A. Simonne, J. Bartz. 2002b. Fresh Produce Handling, Sanitation and Safety Measures: Strawberry, raspberry, blackberry and blueberry. FSHN0212/FS093. Gainesville: University of Florida Institute of Food and Agricultural Sciences. http://ufdc.ufl.edu/l/IR00002057/00001

McAvoy, E.J., N.S. Boyd, M. Ozores-Hampton, P.D. Roberts, J.W. Noling, and H.A. Smith. 2017. "Chapter 8. Eggplant Production." Vegetable Production Handbook. HS726. Gainesville: University of Florida Institute of Food and Agricultural Sciences. http://edis.ifas.ufl.edu/cv124

Miller, C.F., Q. Wang, P.J. Dittmar, E.J. McAvoy, M. Ozores-Hampton, P. Stansley, J.A. Smith, R.N. Raid, C.A. Snodgrass, S.E. Webb, A.J. Whidden, and S. Zhang. 2017. "Chapter 10. Minor Vegetable Crop Production." Vegetable Production Handbook. CV294. Gainesville: University of Florida Institute of Food and Agricultural Sciences. http://edis.ifas.ufl.edu/cv294

Muñoz-Carpena, R. 2004. Field Devices for Monitoring Soil Water Content. BUL343. Gainesville: University of Florida Institute of Food and Agricultural Sciences. http://edis.ifas.ufl.edu/ae266

Muñoz-Carpena, R. and M.D. Dukes. 2005. Automatic Irrigation Based on Soil Moisture for Vegetable Crops. ABE356. Gainesville: University of Florida Institute of Food and Agricultural Sciences. http://edis.ifas.ufl.edu/ae354

Muñoz-Carpena, R. and M.D. Dukes. 2007. Irrigation Virtual Field Day.

Ozores-Hampton, M., P.J. Dittmar, E.J. McAvoy, D.R. Seal, S. Zhang, J.H. Freeman, and Q. Wang. 2017. "Chapter 11. Legume Production." Vegetable Production Handbook. HS727. Gainesville: University of Florida Institute of Food and Agricultural Sciences. http://edis.ifas.ufl.edu/cv125

Ozores-Hampton, M., N.S. Boyd, E.J. McAvoy, C.F. Miller, J.W. Noling, G.E. Vallad. 2017. "Chapter 13. Pepper Production" Vegetable Production Handbook. HS732. Gainesville: University of Florida Institute of Food and Agricultural Sciences. http://edis.ifas.ufl.edu/cv130

Ozores-Hampton, M., R. Kinessary, E.J. McAvoy, R.N. Raid, and J. Beuzelin. 2017. "Chapter 17. Sweet Corn Production." Vegetable Production Handbook. HS737. Gainesville: University of Florida Institute of Food and Agricultural Sciences. http://edis.ifas.ufl.edu/cv135

Runyan, C., T. Obreza, T. Tyson, B. Goodman, P. Tacker, R.Yager, J. Thomas, A. Johnson, G. Grabow, B. Smith, and S. Dennis. 2007. "Maintenance Guide for Microirrigation

Systems in the Southern Region." SR Regional Water Program. http://fawn.ifas.ufl.edu/tools/irrigation/citrus/maintenance/

Simonne, A., M.E. Swisher, K.N. Moore and K. Sattanno. 2017. Small Farm Food Safety, Fresh Produce – Part 6: Additional resources for participants bringing food safety concepts to farms. FCS8850. Gainesville: University of Florida Institute of Food and Agricultural Sciences. http://edis.ifas.ufl.edu/fy971

Simonne, A. 2007. Principles and Practices of Food Safety for Vegetable Production in Florida. FCS8817. Gainesville: University of Florida Institute of Food and Agricultural Sciences.

Simonne, A., J. Brecht, S. Sargent, M. Ritenour, and K. Schneider. 2005. Good worker health and hygiene practices: Training manual for produce handlers. FCS8769. Gainesville: University of Florida Institute of Food and Agricultural Sciences. http://ufdc.ufl.edu/l/IR00002283/00001

Simonne, E.H., D.W. Studstill, R.C. Hochmuth, J.T. Jones and C.W. Starling. 2005. On-farm Demonstration of Soil Water Movement in Vegetables Grown with Plasticulture. HS1008. Gainesville: University of Florida Institute of Food and Agricultural Sciences.

Simonne, E.H., D.W. Studstill, R.C. Hochmuth, T. Olczyk, M. Dukes, R. Munoz-Carpena, and Y. Li. 2017. Drip Irrigation: The BPM Era – an Integrated Approach to Water and Fertilizer Management for Vegetables Grown with Plasticulture. HS917. Gainesville: University of Florida Institute of Food and Agricultural Sciences. http://edis.ifas.ufl.edu/hs172

Smajtrla, A.G. 2005. Hose Connection Vacuum Breakers for Home Backflow Prevention. AE258/AE113. Gainesville: University of Florida Institute of Food and Agricultural Sciences. http://ufdc.ufl.edu//IR00001505/00001

Smajtrla, A.G. and B. Boman. 2002. Flushing Procedures for Microirrigation Systems. BUL333. Gainesville: University of Florida Institute of Food and Agricultural Sciences.

Smajstrla, A.G. and F.S. Zazueta. 2002. Estimating Crop Irrigation Requirements for Irrigation System Design and Consumptive Use Permitting. AE257/AE078. Gainesville: University of Florida Institute of Food and Agricultural Sciences. http://ufdc.ufl.edu//IR00003716/00001

Smajstrla, A.G., B.J. Boman, G.A. Clark, D.Z. Haman, D.S. Harrison, F.T. Izuno, D.J. Pitts and F.S. Zazueta. 2002a. Efficiencies of Florida Agricultural Irrigation Systems. BUL247/AE110. Gainesville: University of Florida Institute of Food and Agricultural Sciences.

Smajstrla, A.G., B.F. Castro and G.A. Clark. 2002b. Energy Requirement for Drip Irrigation of Tomatoes in North Florida. BUL289/AE044. Gainesville: University of Florida Institute of Food and Agricultural Sciences.

Smajstrla, A.G., D.S. Harrison, and J.M. Stanley. 2005. Evaluating Irrigation Pumping Systems. AE24/AE122. Gainesville: University of Florida Institute of Food and Agricultural Sciences. http://ufdc.ufl.edu//IR00001506/00001

Smajstrla, A.G., B.J. Boman, D.Z. Haman, F.T. Izuno, D.J. Pitts and F.S. Zazueta. 2006. Basic Irrigation Scheduling in Florida. BUL249/AE111. Gainesville: University of Florida Institute of Food and Agricultural Sciences. http://ufdc.ufl.edu//IR00001504/00001

Treadwell, D.D. and J. Perez. 2017. Introduction to Organic Vegetable Production. HS720. Gainesville: University of Florida Institute of Food and Agricultural Sciences. http://edis.ifas.ufl.edu/cv118

Vallad, G.E., H.A. Smith, P.J. Dittmar, and J.H. Freeman. 2017. Vegetable Production Handbook of Florida, 2017–2018. CV292. Gainesville: University of Florida Institute of Food and Agricultural Sciences. http://edis.ifas.ufl.edu/cv292

Whitaker, V.M., N.S. Boyd, N.A. Peres, J.W. Noling, and J. Renkema. 2017. "Chapter 16. Strawberry Production." Vegetable Production Handbook. HS736. Gainesville: University of Florida Institute of Food and Agricultural Sciences. http://edis.ifas.ufl.edu/cv134

Zazueta, F.S., A.G. Smajstrla and G.A. Clark. 2002. Irrigation System Controllers. SSAGE22/AE077. Gainesville: University of Florida Institute of Food and Agricultural Sciences. http://ufdc.ufl.edu//IR00001497/00001

Zotarelli, L. P.J. Dittmar, M. Ozores-Hampton, N.S. Dufault, B. Wells, J.W. Noling, E.J. McAvoy, Q. Wang, and C.F. Miller. 2017. "Chapter 6. Cole Crop Production." Vegetable Production Handbook. HS724. Gainesville: University of Florida Institute of Food and Agricultural Sciences. http://edis.ifas.ufl.edu/cv122

Zotarelli, L., P.J. Dittmar, P.D. Roberts, J.W. Noling, and B. Wells. 2017. "Chapter 14. Potato Production." Vegetable Production Handbook. HS733. Gainesville: University of Florida Institute of Food and Agricultural Sciences. http://edis.ifas.ufl.edu/cv131

Zotarelli, L., M. Dukes, G. Liu, E.H. Simonne, and S. Agehara. 2017. "Chapter 3. Principles and Practices of Irrigation Management for Vegetables." Vegetable Production Handbook. AE260. Gainesville: University of Florida Institute of Food and Agricultural Sciences. http://edis.ifas.ufl.edu/cv107

Index